CÉ

CONSTELLATIONS AND CONJECTURES

# SYNTHESE LIBRARY

MONOGRAPHS ON EPISTEMOLOGY,

LOGIC, METHODOLOGY, PHILOSOPHY OF SCIENCE,

SOCIOLOGY OF SCIENCE AND OF KNOWLEDGE,

AND ON THE MATHEMATICAL METHODS OF

SOCIAL AND BEHAVIORAL SCIENCES

*Editors:*

DONALD DAVIDSON, *Rockefeller University and Princeton University*

JAAKKO HINTIKKA, *Academy of Finland and Stanford University*

GABRIËL NUCHELMANS, *University of Leyden*

WESLEY C. SALMON, *Indiana University*

NORWOOD RUSSELL HANSON

# CONSTELLATIONS
# AND CONJECTURES

*Edited by*

WILLARD C. HUMPHREYS, Jr.
*Evergreen State College, Olympia, Washington*

D. REIDEL PUBLISHING COMPANY
DORDRECHT-HOLLAND / BOSTON-U.S.A.

Library of Congress Catalog Card Number 70–159654

ISBN 90 277 0192 X

397463

Published by D. Reidel Publishing Company,
P.O. Box 17, Dordrecht, Holland

Sold and distributed in the U.S.A., Canada, and Mexico
by D. Reidel Publishing Company, Inc.
306 Dartmouth Street, Boston,
Mass. 02116, U.S.A.

Printed in The Netherlands by D. Reidel, Dordrecht

Dare sound Authority confess
That one may err his way to riches
Win glory by mistake, his dear
Through sheer wrong-headedness?

W. H. AUDEN
*The History of Science*

# PREFATORY NOTE

This is the final volume of Norwood Russell Hanson's posthumous works to reach publication, and it forms a companion to the collection of papers, *What I do not Believe, and other Essays*, prepared for the Synthese Library by Harry Woolf and myself.

A word of explanation is necessary about the date and condition of the MSS from which this text has been prepared. Originally, Hanson planned a substantially longer book, which would have brought the story he tells here at least up to the late 19th century, and would have incorporated (e.g.) his well-known paper on the theories of Leverrier. The preliminary work on this larger book was largely completed by 1964, but it was then set aside. Between 1964 and the time of his death in 1967, Hanson worked over the material from the Greeks up to Kepler, and cast it into the new shape it has here. This involved a great deal of revision, and the typescript on which he was working bears very extensive manuscript amendations throughout all but the last few pages. So far as it is possible to tell, accordingly, the book in its present form represents Hanson's final views.

None the less, it is important to point out that the historical material on which the present text was based is substantially that available to the author in the early 1960's. Since the central argument of the book is a philosophical one about the nature of explanation, Professor Humphreys and I decided that it was preferable to leave the text as it stands, with occasional Editor's footnotes (see, e.g., page 5), rather than chop and change Hanson's work in order to bring it into line with the results of the most recent historical scholarship.

The original MS lacked any definitive title, and we have taken the liberty of inventing one, with the concurrence of the General Editor of the Synthese Library, Professor Hintikka.

In conclusion, let me add a personal word of appreciation about the vast amount of patient hard work that Professor Willard C. Humphreys Jr., who was formerly one of Hanson's own students, has devoted to the task of preparing this complex manuscript for the publisher.

*Santa Cruz, California*
*October, 1972*

STEPHEN E. TOULMIN

# CONTENTS

# BOOK TWO – PART II

# BOOK THREE – PART I

# BOOK ONE

## PART I

## Cosmological Explanation, B.C.

# THE CONCEPTUAL CONTENT
# OF BOOK ONE, PART I

*'Hempel's Hypothesis' Concerning Explanation and Prediction*

> Prediction of $x$ is explaining $x$ before it happens
> Explanation of $x$ is predicting $x$ after it happens:
>> there is a special logical symmetry between the concepts of
>> explanation and prediction

*Objections:*

> There are predictions without corresponding explanations
> There are explanations without corresponding predictions

*The History of Planetary Theory is an Interplay Between*
*Predictions* sans *Explanation and Explanations*
sans *Prediction*

> Hempel's hypothesis is realized only briefly in the 17th
> century

# INTRODUCTION

An occurrence is explained by being related to prior events through known laws. Other intellectual activities may also constitute explanation – but this much certainly does. *Ideally*, an explained occurrence (O) could have been predicted in a connected way – by extrapolation from prior events (e) *via* the same laws (L).

Schematically,

$$\textit{Explanation:} \quad O^t \quad -L_{1,\,2,\,3} \rightarrow (e_1 e_2 e_3)^{t-\varDelta t}$$
$$\textit{Prediction:} \quad (e_1 e_2 e_3)^t - L_{1,\,2,\,3} \rightarrow O^{t+\varDelta t}$$

Thus Mars' backward loop in late summer, 1956, is explained by showing how this follows from $(e_1)$ its mean distance from sun and earth, $(e_2)$ its mean period of revolution, $(e_3)$ its past positions relative to earth, etc. – by way of the laws of Celestial Mechanics (including $(L_1)$ Kepler's Laws and Galileo's, $(L_2)$ Newton's, and $(L_3)$ those of Laplace and Lagrange. Moreover, this loop (O) could have been predicted from such events $(e_1 \rightarrow e_3)$ *via* the laws of Celestial Mechanics.

This is an ideal situation. It crystallized late in the history of planetary theory. The Greeks found explanations for heavenly motions: the backward loops were explained to their satisfaction. But they could not predict these motions, not in terms of Attic explanatory cosmologies. Later, Ptolemy *could*, with unprecedented accuracy. But he could not explain them, not via technical astronomy. Ptolemy the Aristotelian *cosmologist* must be distinguished from Ptolemy the geocentric astronomer. Here are two different thinkers united in the same historical person. The cosmologist-Ptolemy repeated verbatim the word-pictures of antiquity whenever he discussed his 'philosophy of the universe'.*

---

* *Editor's note:* A highly important piece of recent scholarship which Hanson would certainly have wished to take into account in this connection is Bernard Goldstein's translation of 'The Arabic Version of Ptolemy's Planetary Hypothesis', *Transactions of the American Philosophical Society*, New Series, **57**, Part 4, 1967. Goldstein has discovered that Ptolemy did in fact undertake to synthesize Aristotelian cosmology and technical astronomy

But the astronomer Ptolemy denies that full explanation of the planetary perturbations lies within human powers. Astronomical explanation is thus virtually inconceivable for Claudius Ptolemy. He settles for *mere* prediction.

Explanation becomes important with Copernicus, more important still for Kepler, and finally culminates in Newton where the Hempelian 'symmetry' of explanation and prediction is instantiated in the ideal manner just schematized. Here was *the* consumately rational scientific system; *The Principia*! Then came Leverrier.

Fully to delineate the interplay between these two currents within the history of planetary theory is a primary purpose of this book. This will require analytical discussions of concepts like *cosmology* and *astronomy*, the *scientific system, 'clean' mathematical ideas, hypothesis-discovering* and *extrapolation*. These themes are fundamental to all the sciences; nothing brings them out with greater clarity and definition than the development of planetary theory.

---

into a coherent system but that this aspect of his work has largely been ignored since the zenith of medieval Arabic science. What was known of this synthesis among Latin writers, however, is not yet clear. Hanson reverts to the point he is making here frequently in what follows and the reader should bear in mind that he is speaking of Ptolemy's well-known *Almagest* and Part I of *Planetary Hypotheses*, not the recently recovered materials which Goldstein has brought to light. Regardless of the historical development, the logical point Hanson is seeking to make about the nature of systematic explanation in planetary theory is well worth making. Certainly the majority of thinkers and writers on the subject of astronomy down to the time of Copernicus separated astronomical calculation and cosmology in just the way Hanson here ascribes to Ptolemy.

# THE HISTORICAL CONTENT OF BOOK ONE, PART I

## The First Great Fact of the Heavens

    The diurnal path of the sun
    The noctural path of the moon and stars
    The circumpolar stars

## The Motion of the Sun: the Ecliptic

    The two explanations of the first fact of the heavens
    A spinning earth, or a spinning sky?
    Anaximander
    Anaximenes
    Xenophanes
    Parmenides
    Empedocles

## The Second Great Fact of the Heavens

    Empedocles
    Pythagoras
    Democritus
    Philolaus

To understand the past adequately, we must forget the present. This advice is easily advanced. It is difficult to heed. Yet the effort is worthwhile; historical and philosophical research withers beneath the intense rays of hindsight. Resist temptations to treat ancient problems as if they required our modern answers! To understand older perplexities fully we must *have* them: we must make ourselves have them. Only then will it be clear why the ancients' answers seemed plausible and reasonable to them; they were men, after all, no less quick-witted than ourselves.

Imagine yourself at sea a few thousand years ago. A good haul of fish aboard, your skiff skims slowly towards the darkening shore; a rare moment in the struggle for existence! It is a time to contemplate the twinkling heavens. Though allowed little leisure to reflect on them, the stars would have become a great natural fact for you. The dawn, the arching circular flight of Phoebus at noon, and then his fiery plunge into the western sea; similarly the nightly spin of the stars around the celestial pole would be familiar. These will be the dependable experiences of your life. They were that for the pre-Pre-Socratics.

Here then is the first great fact of the heavens. All ancient observers were rightly impressed with the observations of the sun, stars and moon moving over the earth in their perfect circles.

"... moving over the earth in perfect circles"; what more natural way to describe these observations? In such a context what could be more obvious than that the stars revolve around some celestial axis?[1]

What *could* make one seek for anything more complex? This very

[1] "Since the fixed stars are always seen to rise from the same place and to set at the same place ... and these stars in their courses from rising to setting remain always at the same distances from one another, while this can only happen with objects moving with circular motion, when the eye (of the observer) is equally distant in all directions from the circumference ... we must assume that the (fixed) stars move circularly, and are fastened in one body, while the eye is equidistant from the circumference of the circles. But a certain star is seen between the Bears which does not change from place to place, but turns about in the position where it is. And since this star appears to be equidistant in all directions from the circumference of the circles in which the rest of the stars move, we must assume that

Fig. 1.   The sun's path from dawn until noon.

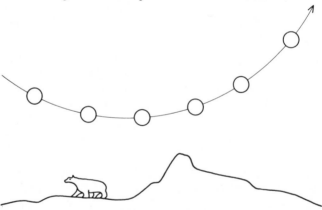

Fig. 2.   Sun's 'nightly' path over Greenland.

question opens the door into the entire history of planetary theory, a door through which we now pass.

The contemplation of the heavens is, hence, a phenomenon of great antiquity. Early in our history the sky was divided into (arbitrary) areas, each embracing a 'constellation' – some 'natural' cluster of celestial lights. It was soon noted that the sun's risings and settings did not

the circles are all parallel, so that the fixed stars move in parallel circles having for one pole the aforesaid star." [Euclid, *Phaenomena* (tr. Heath)].

Fig. 3.　Tracks of the circumpolar stars.

always occur against the same background of stars; rather, it crept along through a band of positions traced among the constellations. This ribbon of positions was called the 'ecliptic'; eclipses were seen to occur only when the moon crossed through it. Understandably, the sun was regarded as a kind of Deity. Each of the constellations it encountered along the ecliptic was animated – being represented mythologically as one of the twelve obstacles, or labors, to be overcome by the solar god. The totality of creatures signified by the constellations was called the 'zodiac', from the Greek *zoon* (a living thing). Hercules was one of the mythological figures associated with the sun; his twelve labors probably derive from descriptions of the zodiacal events just alluded to.

Near the Euphrates have been discovered tablets, dating from 600 B.C.; these name the constellations much as the Greeks of that same period had done. But the ideas behind these names are much more ancient than the tablets themselves. This is apparent from one striking fact. The constellation which we call 'Taurus' (♉) is referred to on the Euphratean tablets as 'Bull-in-front'. The year's beginning was reckoned in those times from the start of spring – the vernal equinox.[2]

Today the sun is in the constellation Pisces (✵) at the vernal equinox. In Hipparchus' time (140 B.C.) it was in Aries (♈). 'Bull-in-front' suggests that when its name was thus given to it, Taurus – the constellation next to the east (before Aries) – contained the vernal equinox. But we

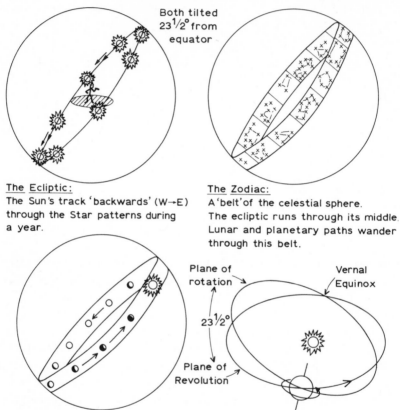

Both tilted 23½° from equator

The Ecliptic:
The Sun's track 'backwards' (W→E) through the Star patterns during a year.

The Zodiac:
A 'belt' of the celestial sphere.
The ecliptic runs through its middle.
Lunar and planetary paths wander through this belt.

Plane of rotation          Vernal Equinox
23½°
Plane of Revolution

Zodiacal belt with positions of moon, in various phases, in course of month.

The Earth at its Autumnal Equinox

Fig. 4.   Ecliptic, Zodiac, and Equinoxes.

[2] In our language, the equinoxes result from the axis of our earth's rotation being in-inclined 23.5° to the ecliptic. So the celestial equator (a skyward projection of our own equator) and the ecliptic, are inclined to each other 23.5°. The two points of intersection are the vernal and autumnal equinoxes. Here day and night are of equal length. (Cf. Figure 4.)

know from calculation that this must have been in 2450 B.C. This is 2000 years earlier than the Euphratean tablets themselves! Doubtless the tablets record a tradition, instead of then-contemporary observations.

Similarly Virgil, captured by this tradition, wrote

The gleaming Bull opens the year with golden horns, and the Dog sinks low, his star (Sirius) averted.

The geometry of the heavens was generally known by Homer's time:

No sleep fell on (Odysseus') eyes; but he watched the Pleiades and the late-setting Wagoner, and the Bear ... which wheels round and round where it is, watching Orion and alone of them all never bathes in the ocean.[3]

Similarly Hesiod:

When the Pleiades, the Hyades and the strength of Orion set, then be mindful of timely ploughing ... But when Orion and Sirius are come into mid-heaven and the rosy-fingered dawn sees Arcturus, then ... cut off all the grape clusters and bring them home ...[4]

Alcaeus:

... the star is coming 'round again, the season is hard to bear with the world athirst ... because Sirius parches head and knees.

Sappho:

The moon is gone and the Pleiads set ...

Theognis:

Foolish are men ... who drink not wine when the Dog-star rules.

Pindar:

And meet it is that Orion should not move far behind the seven mountain Pleiades.[5]

And Simonides wrote of

Atlas' violet-tressed daughters dear, that are called the heavenly Pleiades.

Aratus was the most systematic of the ancient stellar versifiers. His *Phenomena* are referred to often in Cicero:

"... Aratus had illustrated (Eudoxos' globe of the sky) in his verses ...". "The axis is always firm, although it may appear to shift a little, while the earth maintains its equilibrium in the centre, and around it the sky turns itself ... Surrounding (the northerly pole) the two Bears lie circularly ..."

Thus awareness of the first great fact of the heavens is of ancient origin indeed. Many of the ancients realized that for even such striking celestial

[3] Odyssey, V, 270–276.
[4] Works and Days, 615–617, 609–611.
[5] Nemean Odes, ii, 17–18.

data alternative explanations could be offered. The diurnal movement of the sun and noctural motions of the stars *could* be explained by supposing the earth to rotate on its own axis. This rather than the heavens whirling around us – the more usual account, B.C.[6]

[6] This 'explanation' differs in form from the Hempelian paradigm (schematized on page 5). The *explicandum* here is the diurnal-nocturnal motion of heaven. The *explicans* is just the assumption – *not* the further empirical observation – that the earth rotates (or, the counter-explanation, that the heavens rotate). The formula $O^t - L \rightarrow (e_1 e_2 e_3)^{t-\Delta t}$ seems not exemplified here. The '$e_1 \rightarrow e_3$' factual element is lacking.

We have here something like William Dray's analysis [cf. *Explanation in History* (Oxford)], in which $O^t$ is felt to be explained if $e_1 \rightarrow e_3$ are taken to constitute *possible* occurrences – and then *assumed* to obtain. *Why* assumed to obtain? Because if so, O ceases being anomalous. (The neutrino discovery, and that of the planet Neptune, are perfect illustrations of this.) "The diurnal-nocturnal motion of heaven does not designate a unique, particular occurrence, as did 'Mars' backward loop in late summer 1956", or as would "the observed motions of Orion, the Pleiades and Aldebaran between 9 and 11 p.m., 14 July 1966, as from midwestern U.S.A.". These latter describe occurrences like the $O^t$ in our Hempel hypothesis'; but here now O (apparent heavenly motions) is directly explained by L (the *principle* of the earth's rotation). What corresponds to the $e_1 e_2 e_3$ of page 5? Perhaps nothing. Perhaps for some ancients the motions of stars and sun were explained *simpliciter* by L, the assumed rotation of the earth – or *per contra*, the assumed revolution of the heavens.

Thus while noting our first fact of the heavens, we notice also a first complexity of any too-simple analysis of explanation and prediction. It is natural to say that e.g., Hipparchus *explained* the diurnal motions of the stars by assuming the earth to rotate.

Our initial account of scientific explanation (page 5) then, was called 'prediction of the present'; this we characterized as 'covering-law-explanation-of-present-anomalies-by-past-facts'. Thus the standard Hempel analysis of *explanation*. But the rotation-revolution example before us now seems more like an 'explanation-of-present-facts-by-supposition-of-possible-but-unestablished-data'. The cases differ conceptually. One is reminded of Peirce's confusion of two senses of 'retroduction'; (a) the explanation of anomalies by 'reasoning them underneath' covering *laws*, and (b) the explanation of anomalies by 'invoking hidden *objects*' as required by covering laws. Thus, (a) Mercury's perturbations were once thought accounted for by working them into a Celestial Mechanics covered by a gravitation law of the form $F = G^{m_1 m_2}/r^{2+\delta}$ and (b) those perturbations were also once thought explained by reference to a hidden planetoid ('Vulcan') gravitationally required by the unadjusted form $F = G^{m_1 m_2}/r^2$. These two kinds of explanation are different, then. But they are alike, however, in being structurally symmetrical each with their corresponding predictions. It would not be misleading to say that, *because* one day he supposed the earth to spin, Hipparchus predicted that the Bear would appear to turn from East to West that very night. That would be quite different from predicting this inasmuch as the Bear has always appeared thus. The complex conceptual structure of *explanation* and *prediction* will appear in increasing detail as we proceed. But general philosophical remarks belong at the end of an inquiry such as this, not at its beginning. They should be conclusions, not *ex cathedra* preambles. And they belong at the terminus only after detailed descriptive inquiry justifies them, a situation not too often realized in philosophical analyses of *explanation*.

So it was known long ago that there were at least two possible explanations of the nightly revolutions of the circumpolar stars, and of the sun's arching flight over us. The first consisted in supposing that it was the sun and stars which moved around us, here at the 'centre'. (Cf. Figure 5.)

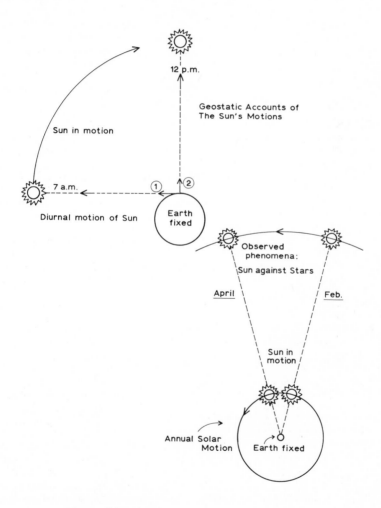

Fig. 5.   Diurnal and annual solar motion: Geostatic.

The second explanation consisted of course in supposing that the earth spins on *its* axis (cf. Figure 6 below).

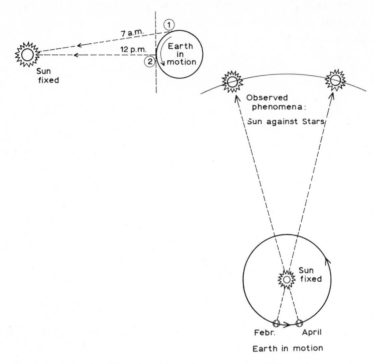

Fig. 6.   Diurnal and annual solar motion: Heliostatic.

Philolaus, Hiketas, Ekphantus, perhaps Plato, and certainly Herakleides stand out among the ancients as philosophers who choose something like this second explanation. Thus Cicero says:

Hiketas of Syracuse ... believes that the heavens, the sun, moon, stars, and all heavenly bodies are standing still, and that nothing in the universe is moving except the earth, which, while it turns and twists itself with the greatest velocity round its axis, produces all the same phenomena as if the heavens were moved and the earth were standing still.[7]

[7] *Acad. prior.*, lib. II. 39, 123: "Hicetas Syracusius... caelum, solem, lunam, stellas, supera denique omnia stare censet, neque praeter terram rem ullam in mundo moveri: quae quum circa axem se summa celeritate convertat et torqueat, eadem effici omnia, quae si, stante terra, caelum moveretur." (Atque hoc etiam Platonem in Timaeo dicere quidam arbitrantur, sed paulo obscurius.)

Aetius says:

Herakleides of Pontus and Ekphantus the Pythagorean let the earth move, not progressively, but in a turning manner like a wheel fitted with an axis, from west to east round its own centre.[8]

Later, Simplicius wrote:

[The Pythagoreans] called the earth a star because it also is an instrument of time, for it is the cause of days and nights, for it makes day to the part illuminated by the sun, but night to the part which is in the cone of the shadow.[9]

He says also:

Herakleides of Pontus and Aristarchus ... thought that the phenomena could be accounted for by supposing the heaven and stars to be at rest, and the earth to be in motion about the poles of the equator from west [to east], making approximately one complete rotation each day.[10]

Further details of Pythagorean cosmology will be set out later.

Despite the boldness of these three ancient philosophers, the idea of a spinning earth seemed conceptually too intricate to most, and observationally unfounded. It was known that the earth was an *immense* object.[11]

Rather obvious effects of spinning ought to be observed here on the surface of so large a sphere. We would indeed be twisting through more than 24 000 miles in less than 24 hours – an absolute speed of more than 1000 mph! If one can 'feel' the effects of running at 10 mph, or riding at 20 mph – experiences familiar to all Greeks of the heroic-philosophical era – how much more ought one to 'feel' a speed of 1000 mph. From

[8] Aet. (Stobaeus), Diels, III, 13, p. 378; Compare also Hippolytus (Philos, XV, Diels, p. 566), who writes:
> "Ekphantus a Syracusan said ... that the earth, the centre of the universe, moved about its own centre towards the east"; Diogenes Laertius records that "Philolaus was the first to hold that the earth rotates in a circle, though others say it was Hiketas of Syracuse". (VIII. 85)

[9] *De Caelo* (Heib.), p. 512.

[10] *De Caelo* II, pp. 444-45 (Heiberg).

[11] Archytas of Tarentum is mentioned by Horace as an earth-measurer (Carm. I. 28). Aristophanes also refers to geometrical instruments used in measuring the whole earth. Aristotle gives the earth's circumference as 400 000 stadia – one stade, or stadium, being 516.73 feet (Pliny, II 247). Archimedes puts the figure at 300 000 stadia (Arenar. I. 8), a value also adopted by Posidonius (Cleom. I. 8, ed. Ziegler, p. 78). Eratosthenes argued for 252 000 stadia (Berger, *Die geographischen Fragmente des Eratosthenes*, Leipzig, 1880), which gives 24 662 miles for the circumference, very close to our modern value (see Celomedes' account of Eratosthenes' methods, *On the Orbits of the Heavenly Bodies* I, 10, translated in T. L. Heath, *Greek Astronomy*).

this it seemed to most thinkers of the period wholly reasonable to conclude that the earth does *not* spin on its axis; the light, wispy and sparkl-
ing heavens really spin around us. *That* is certainly the simpler (i.e.
less-taxing), more rational, observationally well-founded hypothesis
in such a context. *Our* best scientific minds of today, were they
transported back to 400 B.C., and supplied only with data then avaible
to the Greeks, would by-and-large have reached the same geostatic and
geocentric conclusions. The geogyrational alternative suggested by
Philolaus, Hiketas, Ekphantus and Herakleides *goes against the facts* as
understood 2400 years ago.

Similarly, why, with respect to the fixed stars, does the sun rise farther
east each day? One *could* explain this by supposing the *earth* to revolve
yearly around the sun. The rate of the earth's revolution then, would
be precisely the rate at which the sun *appears* to move through the ecliptic.
The two explanations of this might be as set out graphically in Figures 5
and 6.

Aristarchus of Samos is renowned for his advocacy of the second mode
of explanation. Of him Archimedes wrote:

His hypotheses are that the fixed stars and the sun remain unmoved, that the earth revolves
about the sun in the circumference of a circle, the sun lying in the middle of the orbit [12],
and that the sphere of the fixed stars ... is so great that the circle in which he supposes the
earth to revolve bears such a proportion to the distance of the fixed stars as the centre of the
sphere bears to its surface. [13]

But again, this second explanation was regarded – and in the historical
context rightly so – as an observationally unjustified and theoretically
unnecessary complication. It clashed with what the ancients actually
saw. It conflicted with their quite reasonable philosophical doctrines
concerning 'natural places' (shortly to be discussed). Contrast the massive, ponderous earth with the nimble lightness of its swiftest fauna –
birds, horses, and the large cats. Think also of the enormous distance of
the sun and the stars. The earth's motion around the sun, to account for
the apparent eastward progress of the sun through the ecliptic, and

---

[12] But compare O. Neugebauer, *Isis* **34** (1942) 6, for a different translation of this last
phrase.
[13] *Arenarius*, Heath's translation, p. 222: see also Plutarch. (*De facie in orbe lunae* 923 A),
where Aristarchus is said to have assumed the earth's rotation as well. Plutarch also cites
Seleucus of Seleucia as a heliocentrist (*Platonic Questions*, 1006C; Aëtius says the same
thing, III, 17, 9).

the coming and passing of the seasons, would have to be faster than that of the swiftest bird. Indeed, it would have to move forward through space at a velocity almost inconceivable for the ancient mind – again, over 1000 miles an hour. Imagine Gibraltar's great rock lurching towards India at such a speed! No bird could stay with it. And that such a revolutionary speed should be *combined* with a comparable speed of rotation – so that at night our absolute velocity would approach 2000 miles per hour (1000 mph revolution to the east, plus 1000 mph rotation in the same direction), and during the day this velocity would approach absolute rest (due to the rotation of our side of the earth counteracting and nullifying the forward speed of the whole earth in revolution) – this appeared preposterous. Nothing like it was ever experienced – no lurching forward at dusk – no obvious quiescence at dawn. Imagine the rock of Gibraltar twisting thus towards the Orient; a bird on the north side would be momentarily stationary over the earth. But when on the south side of the rock the bird would be moving over the earth at 2000 mph – 33 miles per minute! Clouds, birds, seas – and ourselves – would be left hanging in 'empty space' as our convex platform sprang forward at sunset. But we are not in that plight. So the hypothesis is false, by a good 'hypothetico-deductive' inference. So would run the most reasonable of ancient arguments. (Even in Newton's time there were those who must have so argued: "Many are apt to Fansy that if ye earth moved we should feel its motion ..."[14].)

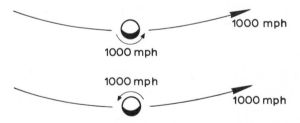

Moreover, if the earth really revolved in a circle around the sun, the diameter of that circle must be immense. Even Aristarchus argues that "The distance of the sun from the earth is greater than eighteen times ... the distance of the moon from the earth".[15] Thus the diametral

[14] Newton, I., *Lord Portsmouth Papers*, 4005, fol. 45–9.
[15] *On the Sizes and Distances of the Sun and the Moon* (translated by Heath, Oxford 1913) Proposition 7.

line across the earth's supposed orbit – from vernal to autumnal equinox, or from winter to summer solstice – must be at least 36 times the distance between the earth and the moon. But just as candles on a dining room wall would group and regroup themselves in different perspectival arrangements as we walk around a central table, so also the stellar constellations should do the same as we move through so vast an absolute distance around the sun. The most careful ancient astronomical observations revealed no such phenomenon.[16] So again, observations available to the ancients seemed simply to disprove the hypothesis of terrestrial motion. The *data* ruled out a twisting, travelling earth.

It is important to realize that the thinkers of antiquity were no less anxious than ourselves to square their ideas with what can actually be observed. Any one of us, put into their historical context, with nothing but their data, ought to have *rejected as empirically unfounded* the complex notion of travelling, twisting earth. Could we have offered reasoning to the contrary any more plausible than that of Aristarchus? As we shall see, Copernicus himself – though his observational and theoretical considerations were more complete – advanced an argument whose logic is identical with Aristarchus', although eighteen centuries separated the two thinkers. If Aristarchus' argument seemed falsified by the observed data, what could have been more reasonable then than to argue that Earth neither rotated nor revolved about the sun? The succession of day and night, the nightly stellar motions and the cycle of the seasons, had to be explained in some other way; some way not so obviously inconsistent with the data of experience.

Philosophers went still further. For some it was not just that the idea of a revolving, rotating earth failed to fit the facts. This hypothesis is more fundamentally untenable; it contradicts the concepts of 'nature' and of 'natural place'. That is, the very notion of a moving earth is internally unsound, given the conceptual matrices which then obtained. This will be explored, believe me.

In short, most *reasonable* men at the time of Homer – and the Old

---

[16] This 'stellar parallax' was not *directly* observed until 1838 when the instruments and painstaking methods of Friedrich Wilhelm Bessel (1784–1846) detected a shift of α Centauri against the background of more distant stars. The shift, of course, resulted from the perspectival relocation of Bessel's telescope, as fixed here on the moving earth, through an absolute distance of 186 000 000 miles – a not-insignificant baseline on which to construct acute celestial triangles.

Testament – would not have entertained any explanation of the first fact of the heavens which made the earth other than immovable. Why should they? They were prudent and wise then to treat their data thus. Their sun and stars, then, simply moved in perfect circles over our immobile earth.

This view is not discussed in the Old Testament – although in *Proverbs*[17] we read: "He set a circle upon the face of the deep". This probably refers to the horizon, however, as in the Homeric poems, where the earth is frequently depicted as a flat disc – again, not an obviously implausible notion.[18]

Anaximander (611–545 B.C.) wrote:

The earth is poised aloft, supported by nothing, and remains where it is because of its equidistance from all other things.[!] Its form is rounded, circular, like a stone pillar; of its plane surfaces one is that on which we stand, the other is opposite.[19]

Anaximander viewed the heavens as spherical, enclosing our atmosphere "as does the bark on a tree"; forming a series of layers, the sun, moon and stars being situated between these – the last mentioned being nearest to us, and the sun farthest.[20]

Eudemus, indeed, cites Anaximander as the first to attempt an ordering of the celestial bodies with respect to distance.[21]

Anaximenes (6th century B.C.) had the firmament turn around the earth, without question. He took the celestial vault to be made of solid, crystalline stuff[22], adding that besides stars it carried some bodies of a terrestrial nature.[23]

Xenophanes (570 B.C. – 5th century) is remarkable for his dissent from

---

[17] viii, 27.
[18] Krates of Mallus (2nd century B.C.) asserts that both Homer and Hesiod knew that the earth was a sphere: Cf. Susemihl, *Gesch. d. griech. Lit. in der Alexanderinerzeit*, II, p. 5.
[19] Hippolytus, *Refutation of all Heresies*, I.6.3, Heath's translation: also Diels p. 559, p. 218. Diogenes (11.1) says Anaximander taught of a spherical earth; but he must be wrong. Aristotle would have mentioned such a signal view when discussing Anaximander's ideas about the earth's equilibrium (*De Caelo* II. 13, p. 295b); he does not. Theon (ed. Martin) p. 324 credits Anaximander, on the authority of Eudemus, with making the earth move "... round the centre of the world". But Aetius and Aristotle pass this over, so the allusion is dubious. (Act. III. 10, p. 376.)
[20] Strom. II Diels, p. 597; Act. II, 15, p. 345; Hippol. Phil. VI, p. 560.
[21] See Simplicius' *Commentary on De Caelo* (Heiberg) II, 10, p. 471.
[22] Aetius II, 14, p. 344.
[23] Hippolytus. Philos. VII; and compare Theon (Martin) p. 324.

the 'majority' opinion that circular motion was the first rule of the cosmos. This poet took the heroic course of arguing that the apparent circularity of celestial movement was an illusion due to the immense distances of sun and stars[24], their *real* motion was rectilinear! But as with poets before and since, Xenophanes' speculations failed to stimulate scientific opinion. Certainly his view did not fit easily with the available observations: it is not an obvious explanation of 6th century (B.C.) data even back then. And yet, how like our contemporary conception of an expanding universe, in which all major motions are traced rectilinearly away from an original point of immense potential energy.

Something called 'the existent' was a fundamental philosophical consideration for Parmenides (5th century B.C.). It is "perfect on every side, like the mass of a rounded sphere, equidistant from the centre at every point", and continuous.[25]

Parmenides' universe is a system of concentric spheres. So the earth, he argues[26], ought to be of the same figure; a singular argument.[27]

This is the first suggestion of a concentrically spherical cosmos; it comes to have a fruitful future in astronomical explanation. Outermost was the sphere of the morning star and the evening star (which Parmenides knew to be one and the same). Closer yet to earth spins the sphere of the sun and the moon, both being equal in size – phenomenologically a fairly accurate equation[28]. In equilibrium at the centre is earth – which remains at rest *because* its tendency to fall in one direction is no stronger than its tendency to fall in any other direction.[29]

Empedocles' (c. 450 B.C.) universe was finite and spherical. Here, however, we encounter a further idea. Properly to introduce it we must harken back to the as-yet-unmentioned Pythagoras (6th century B.C.).

[24] Aetius II, 24, p. 355.
[25] Fragments. 102–109, also Fairbanks, p. 96.
[26] As did Aristotle, *De Caelo*, II, 4, p. 287a.
[27] Compare Diogenes Laertius, VIII, 48 (comp. ix. 21), where Parmenides is credited as the first to assert the earth's spherical form. This, if true, is doubtless the result of Parmenides' noting circumpolar stars as they become 'risers and setters' during his travels south from Greece, and *vice versa* as he traveled north. Krates of Mallus interprets Homer's *Odyssey* (x. 82ff.) as making this same point (Geminus, vi, 10 (p. 72, Manitius ed.)).
[28] Aetius II, 7 – Diels p. 335.
[29] Aetius III, 15, p. 380; compare Anaximander's first words on p. 21 above. Notice that this last argument is identical with what was later called 'The Principle of Indifference'. It appeared in Archimedes' demonstration of the law of moments. (*De Aequiponderantibus*),

Because, if Theon of Smyrna is correct, Pythagoras was the discoverer of the *second* great fact of the heavens, a fact so difficult to reconcile with the first that the entire subsequent history of planetary theory may be viewed as a succession of efforts to square these two. This second fact is that the planets do not move uniformly with the rest of the heavens. To explain this phenomenon fully required two millennia. Theon wrote:

The impression of variation in the movement of the planets is produced by the fact that they appear to us to be carried through the signs of the zodiac in certain circles of their own, being fastened in spheres of their own and moved by their motion, as Pythagoras was the first to observe, a certain varied and irregular motion being thus grafted, as a qualification, upon their simply and uniformly ordered motion ...[30]

Theon's attribution of this discovery to Pythagoras is apocryphal. Otherwise Aetius would have mentioned Pythagoras when he wrote:

Some mathematicians hold that the planets hold a course opposite to that of the fixed stars, namely from west to east. With this Alcmaeon too, agrees.[31]

Nonetheless, after Pythagoras observations of the west → east motion of the planets began to make the idea of a single spinning celestial sphere (concentric with our fixed earth), difficult to hold in unmodified form. What were these observations like?

This second fact of the heavens is phenomenologically complex. It consists in noting that each evening the planets rise a little farther east against their zodiacal background – and also that sometimes they halt altogether, and even back up! That is, against the simplicity of the *first* fact of the heavens – that each heavenly body circles our immobile earth from east to west every 24 hours – against this must be mounted the fearfully unsettling observation that the planets not only inch along slowly, from west to east, but also that they sometimes loop backwards on their paths. The obvious and natural geostatic explanations of the first fact fail totally with the second.

---

in the tale of Buridan's ass (*Quaestiones in decem libros ethicorum Aristotelis*; Book III, question I, raises the issue, but mentioned no ass; neither does Dante, *Paradiso* IV): and compare LaPlace's *Essai philosophique sur les probabilités* with its 'Principle of The Equal Distribution of Ignorance'.

[30] Heath's transl., pp. 11–12.
[31] II, 16, p. 345.

Put yourself again into the ancients' context. After finding the most fitting explanations of data such as those recorded in Figures 1, 2 and 3, what would one make of data such as these following?

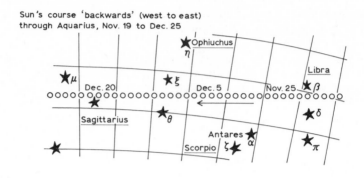

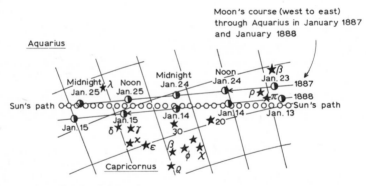

Fig. 7.   Solar and lunar paths through constellations.

And consider Mars and Venus during the years 1958–62 (see Figure 11).

Theon credits Pythagoras with an attempted reconciliation and explanation of these two great facts. But doubtless his 20/20 hindsight, plus his blind veneration of Pythagoras, collapsed several centuries together in his mind. The many explanations of this second fact will be discussed in detail in this book, for they *are* the history of planetary theory. What matters here is that the errant motions of the planets are at last known, perhaps to Pythagoras, certainly to Empedocles.

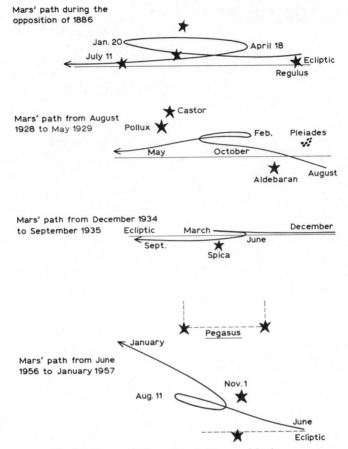

Fig. 8.   Retrogradations: Saturn, Mars, and Jupiter.

Empedocles' sphere of fixed stars is solid and crystalline. And *within that sphere the planets wander freely.*[32]

Here then are *the* two facts of the heavens squarely faced together. For Empedocles, the fixity, internal rigidity and circular motion of the constellations were explicable only if imbedded in a solid, crystal sphere which twists around the earth – a conception which gradually grew in importance. Yet the planets did not partake of this stellar motion.

---

[32] Aetius II, 13, pp, 341–2; it is perhaps unnecessary to note that 'planets' – πλαμετισ – *meant* 'wandering stars'. Thus, even Newton: "A planet in Greek signifies a wandering body". (MS. Add 4005 fols. 45–9).

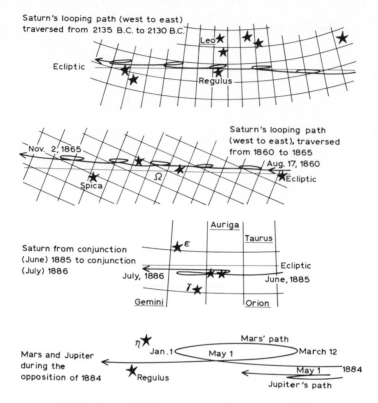

Fig. 9.    Retrogradations: Mars – 1886, 1929, 1935, 1957.

They wandered about like bees in a balloon. Here the history of planetary theory really begins in earnest. Because the ultimate reconciliation of the 'wandering' planetary motions with the inexorably regular motions of the stars – this was shaped only by Newton and polished by Laplace – unwittingly to be shattered later by Leverrier.[33]

So, well before the 5th century B.C., Mercury, Venus, Mars, Jupiter and Saturn had been distinguished from the stars behind them, and had been marked as 'wanderers'. But the anomaly within these conflicting facts did not nag cosmological speculation for some time. The primary fact of the heavens remained its perfectly circular motions. Planetary misbehaviour was awkward, but the aberrations were slow and nothing

[33] Cf. *Supra*, Book Three.

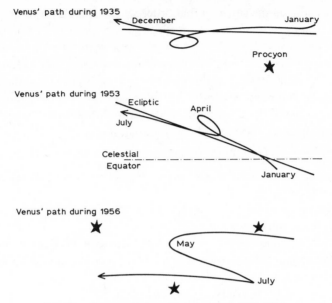

Fig. 10.   Retrogradations: Venus – 1935, 1953, 1956.

like so dramatic as the nightly circuit of the stars around the pole, or the daily vault of the sun from horizon to horizon.

Democritus (5th century B.C.) felt that the earth, being 'heavier', had settled to the centre of the universe. Lighter elements – water, air and

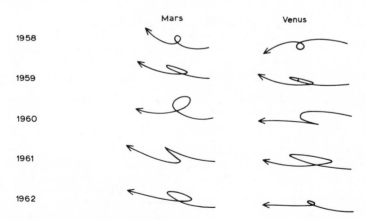

Fig. 11.   Retrogradations: Mars and Venus – 1958→1962.

fire – ensphered the earth, in that order.[34] The planets however, are not singled out by him as erratic Empedoclean wanderers; they simply move from east to west more slowly than the fixed stars.[35]

Anaxagoras (fl. 456 B.C.) ordered the 'planets' thus: Moon (☾), Sun (☉), Mercury (☿), Venus (♀), Mars (♂), Jupiter (♃), and Saturn (♄). Both Plato and Aristotle later adopted this splendid arrangement. But Anaxagoras knew nothing of the orbital motion of these bodies; he mentions only their too-obvious diurnal motion.[36]

Of Anaxagoras, Hippolytus writes:

The sun and the moon and all the stars are fiery stones carried round by the rotation of the aether ... The course of the stars goes under the earth ...[37]

All Pythagoreans took the universe to be spherical. Since the earth is its centre, it too must be spherical.[38]

Aëtius says (see above p. 23) that Alcmaeon of Kroton, and the mathematicians, discovered that "the planets hold a course opposite to

---

[34] Aetius III, 13, p. 378, and Diogenes Laertius, ix 31–33.

[35] (Aetius II, 16, p. 345.) Lucretius also remarks that the bodies nearest to the earth move slowest: *De Rerum Nature*, v. 619–626:

> "So the earth by its weight and the coalescing of its substance came to the rest ... drifted downward and settled at the bottom like dregs. Then sea and air and fiery ether itself were each in turn left unalloyed in their elemental purity, one being lighter than another ... the nearer the heavenly bodies are to the earth, the less they are caught up in the vortex of the heavens. The rushing and impulsive energy of the vortex ... fades out and dwindles at lower levels ..."

Alexander of Aphrodisias credits the same to the Pythagoreans:

> "They said that those bodies move most quickly which move at the greatest distance, that those bodies move most slowly which are at the least distance, and that the bodies at intermediate distances move at speeds corresponding to the sizes of the orbits." [*Commentary on Aristotle's Metaphysics* (Brandis, p. 524a5–18) Heath's translation].

The link between planetary orbital velocity and orbital size, here intuited by Alexander *et. al.*, is finally forged only in Kepler's Third Law, of 1619.

[36] Cf. Proklus, *In Timaeum*, p. 258c, who writes on the authority of the Aristotelian astronomer, Eudemus.

[37] *Refutation of all Heresies*, I.8.6–10, Burnet's translation.

[38] This argument is remarkable again – identical in form to that of Parmenides (above, p. 22) and Aristotle (*De Caelo*, II, 4, p. 287a). Fully to comprehend why the ancients could not designate 'a cube' or 'a pyramid' or 'a tetrahedron' as possible answers to the question 'What is the shape of a spherical universe's centre?', is to grasp the essence of Greek explanatory science. The argument is set out in Diogenes Laertius (viii. 25) and repeated verbatim in Suidas (see πυθαγόρας Σάμιος). But compare W. A. Heidel's *The Frame of*

that of the fixed stars, namely from west to east".[39] This is more reliable than what was asserted earlier by Theon, who made Pythagoras himself the discoverer. So these Pythagoreans differed from the Ionians (and others) in observing not just that the planets move slower than stars, but that they move in the opposite direction. As noted, Philolaus tried to explain the apparent circular motion of sun and stars by a progressive terrestrial revolution in a tiny circle; this is not rotation, but swift revolution in a small orbit.[40]

This proposal might have resolved the tension between the great facts – that stars travel from east to west, and planets from west to east. Because here, with Philolaus, the fixed stars really are fixed. Stellar movement is merely apparent; only the planets move, in large orbits from west to east. This is also the direction in which the earth progresses in its tiny and tight daily orbit. All celestial motion is thus in the same direction, despite appearances. As Aristotle remarks of this Pythagorean cosmos:

The earth, being one of the stars, is carried in a circle round the centre and produces thereby day and night …[41]

A further feature of Philolaus' cosmology is also important; he thought the earth too gross and dross to occupy the absolute centre of all things in the universe. Aristotle gives the reason:

But also many others (besides the Pythagoreans) may share this opinion that one ought not to place the earth in the middle, as they do not take their convictions from the phenomena, but from [rational] considerations (ἐκ τῶν λόγων) – for they think that the most excellent ought to have the most excellent place; but fire is more excellent than earth …[42]

---

the Ancient Greek Maps, for an argument against Pythagoras being the first to assert the earth's sphericity. Theophrastus gives the priority to Parmenides.

Of course, Leukippus and Democritus do, in effect, answer the above question with the remark that the earth is discus-shaped (Aetius III, 10, p. 377), or tympanic (Diogenes Laertius lx. 30). But that this had no effect on subsequent cosmology only proves how much more 'natural' seemed the reasonings of Parmenides, the Pythagoreans and Aristotle.

[39] II. 16, p. 345.

[40] Aëtius II, 4, 5, 7, 20, 30; III, 11. Diels pp. 332b, 333a, b, 336–37b, 349a, b, 361b, 377a. Diogenes Laertius VIII 85: "(Philolaus) was the first to affirm that the earth moves in a circle …". Compare Böckh's Kleine Schriften, vol. III, Berlin 1866.

[41] De Caelo II, 13, p. 293a; Philolaus did not think of the earth as rotating; there was no other case of rotation in the cosmos. The moon, remember, was thought not to rotate because it kept but one face towards the earth all the time (Aristotle, De Caelo, II, 8, p. 290, a.26). Even Kepler argues thus. He explains the moon's lack of rotation as follows: "Gyratio igitur in Luna, ut supervacue, fuit omissa", in Gesammelte Werke, vol. 7, p. 319 (Ed. Caspar).

[42] Ibid.

So Philolaus, Pythagoreans, and contemporaries of Aristotle, displaced earth from the centre of the universe – not for observational reasons, but for theoretical ones. Cosmological explanation here had to be masked with the philosophical distinctions of the Pythagoreans – distinctions concerning the relative excellences of the primitive elements *earth, water, air* and *fire*. That the basic consideration here was theoretical, not observational, is clear from Philolaus' discussion of 'the central fire'. This must not be identified with our sun[43]. The central fire is never seen. Yet 'round this ineffable object earth, sun, moon and planets revolve in perfect circles. The explanatory principle here is simply (and crudely), that *nothing* observable could possibly be perfect enough to reside at the centre of the universe.[44]

Aristotle writes:

... everyone who regards the heavens as finite says that (the earth) lies at the centre. But the Italian philosophers known as Pythagoreans take the contary view. At the centre, they say, is fire, and earth is one of the stars, creating night and day by its circular motion about the centre. They further construct another earth in opposition to ours, to which they give the name 'counter-earth' ... All who deny that the earth lies at the centre think that it revolves about the centre – and not the earth only but ... the counter-earth as well ... there is no more difficulty, they think, in accounting for the observed facts on their view ... than on the common view that the earth is in the middle.[45]

But on this Aristotle muses negatively:

Even as it is, *there is nothing in the observations* to suggest that we are removed from the centre ...[46]

Nonetheless, Philolaus' system remains remarkable in seeking to explain celestial perturbations *via* a theory neither geocentric, nor geostatic, nor heliocentric, nor heliostatic. Earth, sun and a counter-earth *all revolve* around a central fire, always presenting their same hemispheres to that fire. So Mediterranean peoples could see neither The Centre, nor the counter-earth (which Pythagoreans called 'Antic-

---

[43] See Stobaeus, Diels p. 337.

[44] How like much of our history of theological argument; 'no observational predicates are fit to encapsulate the properties of Deity'.

[45] The Pythagoreans held: "that parallax is as negligible in one case as in the other". Heath, *Aristarchus*, pp. 100–101.

[46] De Caelo, II. 13; compare Metaphysics I. 5. p. 986a; my italics – these underline the sentiments of page 21 and 22 *supra*.

thon') – since this latter revolves with the earth, but in a smaller orbit.[47]

Incidentally, Aetius ascribes the Anticthon 'doctrine not to Philolaus, but to Hiketas[48]. Diogenes Laertius reiterates the same suggestion[49]. The tight orbit into which Philolaus twisted the earth is collapsed to zero by Hiketas, leaving the planet 'spinning' on its axis. But this was probably conceived as 'revolution in a zero diameter orbit'. Ekphantus as we saw also let the earth move, rather than the sun and stars[50]. And of Herakleides Simplicius writes "By assuming that the earth was at the centre and rotating while the heaven was at rest, Heraklides of Pontus thought he 'saved the phenomena'." (This is the *apparentias salvare* of medieval science.)

---

[47] Simplicius, *De Caelo*, II. 13, p. 511 (Heib.) and Alexander of Aphrodisias *In Metaphysica* (in Aristotle's *Opera* v. p. 1513 198). The 'unobservability' of *Anticthon* reminds the historian of a similarly-described planetary object later called 'Vulcan', which was invented to explain Mercury's misbehaviour at perihelion. The analogies between 'counter-earth' and 'counter-Mercury' run deep!

[48] Act. III 9, Diels, p. 376.

[49] viii. 85.

[50] Compare Hippolytus, Philos. XV, Diels, p. 566.

## Plato

His universe is spherical, geocentric and bespindled
He never notes planetary orbits as non-parallel, nor does he
question their constant velocity

## The Ordering of The Planets

Intersection of zodiac and equator at 24°

## Plato Notes the Second Fact of the Heavens, but Subordinates it to the First

In the *Phaedrus* Plato (427–347 B.C.) describes the universe as being a sphere. The earth is placed in the centre of the heavens in the *Phaedo*. It has no reason to fall one way rather than another, and *hence* remains fixed in the middle.[51] It is an unsupported sphere, our earth, around which the universe rotates diurnally.[52] The spheres of the fixed stars – and the seven 'planets' – are all set on a celestial spindle. But Plato never notes that the planets fail to move in parallel lines or that their orbits have different inclinations to the ecliptic. He does not even recognize that their orbits are not parallel to the celestial equator; nor does he question whether planets move with uniform velocity. The spindle

must turn round in a circle with the whole (universe) that it carries, and while the whole is turning round, the seven inner circles are slowly turned round in the opposite direction to the whole (i.e. the planets move slowly from west to east, while still joining in the daily celestial motion from east to west.)[52]

The universe is said by Timaeus to have *the* perfect figure, that of a sphere.[53] It turns uniformly on its axis, with no other motion. It does not revolve, therefore. Within the outer sphere are the planets, at distances $1 = ☽, 2 = ☉, 3 = ♀, 4 = ☿, 8 = ♂, 9 = ♃, 27 = ♄$.[54] Two major circles, the celestial equator and the zodiac, divide Plato's universe; these "are separated from each other by the angle subtended by the side of [an inscribed, regular] pentadecagon [i.e., 24°]"[55].

Timaeus is made to speak as follows of these circles:

He (the Creator) split the whole of this composition lengthwise into two halves, laying them across like the letter 'X'. Next he bent these into circles and connected them with themselves

[51] *Phaedo*, pp. 107–110. Again, here is the fascinating argument from indifference.
[52] *Republic*, X. pp. 616–617.
[53] *Timaeus*, p. 34a.
[54] "... the earth, which lies at the centre, is 'rolled', and in motion, about the axis (spindle) of the whole heaven. So it stands written in the Timaeus." (Aristotle, *De Caelo*, II. 13.) Heath declares that Aristotle misrepresents Plato's view here, and that in the *Timaeus* it is nowhere affirmed that the earth has any motion whatever; *Aristarchus of Samos*, pp. 174–178. But compare Cornford, *Plato's Cosmology*, p. 130.
[55] Theon of Smyrnà, p. 198 (Hiller), after Eudemus *Astronomies*.

and with each other, so that their extremities intersected at the point opposite ... He caused [the external] circle (i.e., the celestial equator) to revolve along the side of a parallelogram towards the right, and [the internal circle, i.e., the zodiac] along the diagonal towards the left ... the inner circle he divided into six parts, forming thus seven unequal circles arranged by double and triple intervals (i.e., by 1, 2, 4, 8 and 1, 3, 9, 27, giving 1, 2, 3, 4, 8, 9, 27), three of each [progression]. He bade these circles to move in directions contrary to each other; three at equal velocities (i.e., sun, Mercury, Venus), the others with velocities unequal to each other or to these first three, yet in fixed ratios to their velocities.[56]

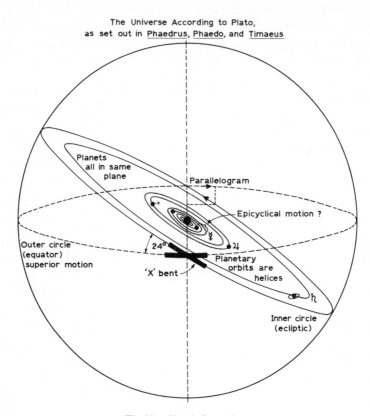

Fig. 12.   Plato's Cosmology.

Plato's question: "By the assumption of what uniform and ordered motions can the apparent motions of the planets be accounted for?"[57]

---

[56] *Timaeus*, p. 36B.
[57] Sosigenes, *De Caelo* (Heiberg), 488.18.

Aristotle in *De Anima*[58] reiterates all this – remarking that the internal circle encloses seven further circles, each one pursuing from west to east its independent circumterrestrial motion, in a plane inclined to the celestial equator. But since the east-to-west motion of the outer circles is 'superior', and over-riding, the diurnal rotation of the celestial equator sweeps the inner machinery along with it. The progress of the internal rings on their reversed paths is thus much slower by comparison.

Plato, therefore, delineates with emphasis the second fact of the heavens. The anomalous behaviour of the planets can no longer be ignored. After Plato, no cosmology or astronomy could be entertained seriously if it failed to note these aberrant motions. But just as Empedocles had fixed on the *major* circular motions in the heavens, neglecting the planets (which he floated freely within the stellar sphere) – so now Plato gives the 'superior' motion to the stars and accommodates the wanderers as best he can. Even so, he moves all the planets in the same plane, an indication that he is not actively seeking further anomalies; rather he's trying to quieten those already upon him.

Timaeus describes the planets and their orbits:

> ... the moon in the first orbit nearest to the earth, the sun in the second above the earth, then the morning star (Venus) and the star sacred to Hermes (Mercury), revolving in their orbits with the same speed as the sun, *but having received a force opposed to it*, owing to which the sun, Mercury and Venus in like manner overtake and are overtaken by each other.[59]

The italics show that Plato construed Mercury and Venus as different from the other planets; this notwithstanding a rival interpretation by Proklus[60]. Theon supposed this passage above to refer to an epicyclical theory, the initial discovery of which he therefore credits to Plato. He is followed in this by Chalcidius.[61] Proklus rejects this; 'Plato nowhere alludes to epicycles of any kind'. It is simply a non-uniform orbital motion of Venus and Mercury, Proklus thought, to which Plato refers. This is unsatisfactory, of course; it is completely 'non-Platonic' that a celestial body should ever move with *other* than perfectly uniform motion! But Plato probably knew little about stationary points and retrograde motions – which weakens the interpretations of both Theon and Proklus.

---

[58] I.3, p. 406B.
[59] *Op. cit.*, p. 38D.
[60] In *Timaeus*, p. 221E.
[61] Cf. Wrobel ed., p. 176; Martin, Theon pp. 302, 424.

He notes only[62] that the courses of the other five planets have been reflected on by but few; their erratic wanderings are infinite in number and of unlimited variety.[63] The diurnal rotation of the heavens twists the planetary orbits (which are inclined to the celestial equator) into helices.[64] So the planets do not, like the stars, describe closed, perfect circles from day to day. Their distances from the equator are always changing.[65]

Clearly then, Plato's cosmology as articulated in the *Timaeus*, is made uneasy by the second fact of the heavens – the meanderings of the planets. The dominant feature in Plato's outlook remains the circular and uniform rotation of the stellar sphere. This is the first great fact; to it Plato subordinates the second – not *vice versa*. That is, his problem is always: *what kind of uniform, circular motion is that of the planets such that they appear to move differently from the fixed stars?* It is never asked (and with good reason) 'what kind of errant motion is that of the stars such that they appear to move in perfect circles, something planets never do?' Indeed, Timaeus remarks "and he made the universe a circle, moving in a circle, one and solitary"[66]; all the world has, essentially, just one motion. Sooner or later all celestial motion must be traceable to the rotation of the stellar sphere around the mathematical line at whose centre the immobile earth is fixed.[67] In the *Epinomis* we are reminded that of the eight 'powers' in the heavens, the fourth and the fifth ( ♀, ☿ ) have the same average velocity as the sun. The eighth moves against all others, drawing them all with it.[68]

Plato rarely goes beyond general statements about the gross features of planetary motion. Still, *Timaeus* indicates that the implications of the second fact of the heavens were making themselves apparent. Not only did the planets move opposite to the direction the stars travelled – orthodox cosmology might have accommodated this – but occasionally they stopped altogether! As we saw, they sometimes looped 'backwards' from east to west. No simple subordination of planetary motion to that of the stars, such as Plato proposed, was going to tame such wanderers; not

---

[62] *Op. cit.*, p. 39C.

[63] *Op. cit.*, p. 39.

[64] *Op. cit.*, pp. 38E–39A.

[65] Cf. Proklus, *op. cit.*, p. 262F, and Theon of Smyrna (ed. Martin), p. 328.

[66] p. 34.

[67] *Timaeus*, p. 40.

[68] 987B.

without radically new mechanical ideas concerning the planet's relations to the stellar sphere. Even though Plato fixed all celestial motions to the rotation of one mighty celestial axle, he did not help us to visualize how, on his hypothesis, the planets could move at different speeds (forward and backwards), and also be inclined at *all* angles to the celestial equator up to 24°. Although well-known before Plato, the necessity and the difficulty of reconciling the two major facts of the heavens are seriously felt only in the *Timaeus*. Before this, cosmologists could speculate on the design of the universe with absolute freedom. Mathematicians and philosophers were now obliged, however, to search out some single *explicans* which would harmonize observations of planets and stars, and set out their mechanical relations in some intelligible order. It was no longer possible simply to recognize planetary misbehaviour and then shrug it off as a temporarily awkward datum. This had now to be faced as *the* central problem of cosmology and astronomy.

*Eudoxos and 'Plato's Problem'*

> The classical arguments for sphericity and geocentricity
> The structure of Eudoxos' theory

*Prediction vs. Explanation*

> Mathematics vs. Physics
> Astronomy vs. Cosmology
> Calendrical computation vs. systematic astronomy

*Intelligibility and the Greek Mind*

On to the scene strides Eudoxos (408–355 B.C.), and with him the history of planetary theory quickens. For, assuming Plato to have posed the basic question for all astronomy, Eudoxos propounded the first serious answer.[69]

Eudoxos is known to mathematicians as the author of the Fifth book of Euclid's *Elements*; he also discovered the 'method of exhaustion', an ancient precursor of our integral calculus. Plato's query, if Simplicius is correct, turned Eudoxos' skills to accounting for the motions of the planets. His was the first substantial attempt to account for those increasingly conspicuous irregularities. Indeed, Eudoxos succeeded to a remarkable degree in representing the major celestial phenomena known in his day.

Most of what we know of Eudoxos' theory comes from Aristotle[70] and Simplicius[71]. The Eudoxian system of concentric spheres elegantly embodies the principle which had become fundamental in astronomy – remained unquestioned for two thousand years: *the planets, however they appear to move, actually traverse perfectly circular orbits.* This, in effect, explains the second fact of the heavens *via* whatever explanation serves for the first fact. It embodies suggestions of all previous thinkers who detected irregularities in the planets, including Timaeus. Circular motion was the only proper and intelligible motion for celestial objects. Nor was this anything but reasonable in the circumstances then. It is not to be dismissed as an irrational prejudice, as is too often done by those im-

---

[69] Thus Simplicius writes:
> "Eudoxos of Cnidus ... is said to have been the first of the Greeks to deal with such an hypothesis. For Plato, Sosigenes says, set this problem for students of astronomy: 'By the assumption of what uniform and ordered motions can the apparent motions of the planets be accounted for?'" (*De Caelo* (Heiberg) p. 488.18).

Plato's question is answered by Newton, but Eudoxos showed the way to all those who preceded the Englishman.

[70] *Metaphysics* (Λ8).

[71] *De Caelo* (II.12, pp. 493–506 (Heib.)).

patient with any idea which does not lay on the path towards modern conceptions.

Do not the stars show us which translation is proper to the heavens? They rise each evening from the same starting point, arch over us in parallel circles, and set in the same place as the night before. During this, their brightness remains constant. This could not be so if they moved in anything but a constant-radius curve, always equidistant from ourselves – the observers. The angular distances between them also remain constant, another consequence of perfectly circular motion. And must we not grant that the circle is the 'easiest' path of motion, free from sudden bends or reversals? Had this not been so some other shape for the chariot-wheel would have forced itself upon us. This is surely the 'natural' form for planetary orbits; consider the spherical and circular forms of earthly things like droplets of rain, and globules of oil in water. Let us anticipate Copernicus here, with some words from *De Revolutionibus orbium Coelestium*:

> ... the universe is spherical in form, partly because this form being a perfect whole, re-quiring no joints, is the most complete of all, partly because it makes the most capacious form ... or again, because all the constituent parts of the universe ... appear in this form; or because everything strives to attain this form, as appears in the case of drops of water and other fluid bodies ... So no one will doubt that this form belongs to the heavenly bodies ...[72]

Indeed, time-measurement depends on sphericity and circularity being the form and motion of the heavens. Thus the argument with most Greek thinkers of Eudoxos' time, and long after.[73]

Moreover, the earth was believed *not* to move. This, for the powerful reason that, if it did, some stellar parallax should be apparent. But the internal geometry of the constellations remains rigid. The stars in the Bear are *never* seen drawing together as they would do if viewed first from one angle and then from another.

If the columns of Hercules 'move' perspectivally as we sail past, how much more should the perspective of the Great Bear shift if we moved as far as necessary to explain (by terrestrial translation) all celestial motions? No shift whatever is observed. Of course, were the fixed stars at an *infinite* distance from earth, this lack of parallax would be (as noted) observationally compatible with an orbital motion for the earth. But, as

[72]  Book I, 1.
[73]  Claudius Ptolemy summarizes all this tersely and beautifully in the *Almagest* (Book I, 3).

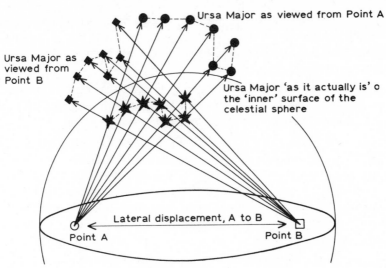

Fig. 13.   Illustration of stellar parallax.

Argument: If the constellations reside in the stellar sphere, and if the earth moves – then the configurations of the constellations should be observed to change. The configurations do not change. The stellar sphere *is* at a finite distance (it revolves around us in a finite time). (So the absence of changes in stellar configurations cannot be attributed to the stars' infinite distance from us.) Hence the earth does not move. *Quod erat demonstrandum.*

Aristotle was soon to argue, the stars cannot be infinitely distant because 'they turn around us' in a finite time; once every twenty-four hours. So they must be at a finite distance. Otherwise their instantaneous velocities would have to be infinite, which we *see* not to be the case: it obviously takes twice as long for Orion to rise and set, as it does for it only to rise to its zenith. Hence its velocity must be finite. Since no parallax is observed, and the stars are only a finite distance away, the earth must be stationary.   Q.E.D.

This argument was a formidable obstacle for all non-geocentric systems. Its plausibility may be appreciated from note 16 where this stellar parallax was remarked to have been observed by Friedrich Wilhelm Bessel (1784–1846) only in 1838. The naked-eye astronomers of 400 B.C. could not even have discerned that Alpha Centauri and 61 Cygni were double stars, much less detect in them (as did Bessel) a parallax relative to more distant constellations.[74]

[74] Yet the logic of Aristotle's argument remains sound. Later helio-centrists kept seeking

Nor is there any variation in the stars' brightness, which there should be if the Earth moved now closer, now farther, from them. And, an unconnected point, all unsupported weights fall toward the center of this spherical earth.[75]

That the immobile earth *must* be spherical seemed clear from the following: (1) the sun and the moon, each rises and sets for different observers at different times, depending on the longitude of observation, (2) eclipses are recorded at different places (east and west) at different times, (3) the earth cannot be concave, or flat, since if this were so the stars would appear above the horizon at intervals different from where they are now seen, (4) nor can the earth be cylindrical, since (if it were) either all the stars would be circumpolar, or none of them would – depending on whether the observer is on the top or the side of the cylinder, (5) and finally, the evidence of ships sailing away to the horizon is incontestable – they descend mast last, no matter what point of the compass they head toward. So the earth is a sphere which has 'fallen to the center of the universe'.[76]

Where could the earth 'fall' to now? It must by its nature remain in the center. Of all elements – fire, air, water, and earth – earth is demonstrably heaviest. Take a bottle and almost fill it with clear water, then add a handful of earth. Rapid shaking of the bottle will mix earth, water, and air evenly until the whole mass seems a spongy, porous mud. But then let the bottle sit. After a week of quiescence, the earth will have settled solidly to the bottom. The water will float on top of this. And the air will have returned to its 'natural layer' above. Thus

---

parallax *because* of that soundness. If the stars are a finite distance away, and *we* move – then parallax must be in principle detectable (as it is!). The alternatives would be (1) to deny heliocentrism or (2) to deny the finite distance of the stars. No post-Newtonian astronomer could have done either thing. So parallax *had* to turn up sooner or later.

[75] This argument is viable even for Newton.

> "... Fixed Stars, Planets, and Comets ... have a gravitating power tending towards them by which their parts fall down ... as stones and other parts of the Earth do here towards the earth and by means of this gravity it is that they are all spherical ..." (MS. Add. 4005, fols. 21–2, Portsmouth Collection).

[76] Such considerations are prominent in Aristotle and Ptolemy, in Copernicus, and even in Newton:

> "... ye earth casts a round Shaddow upon the Moon in Eclipses and by reason of its round figure the Pole star rises and falls as we sail north and south ..." (MS. Add. 4005, fols. 21–2).

the heaviest things fall. And the heaviest things in the universe fall to the center of the universe. So the Earth, being made of earth(!) has no reason to move from where it has fallen to, right here in the center!

We have noted the (plausible) argument that, while light weight bodies should fly off the land (if Earth moved as it should to account for the solar and stellar motions) they do not do so. Leaves fall gently; clouds roam slowly in all directions. Birds fly north, south, east, or west with equal velocity; indeed, the hawk ascends more rapidly than most birds return down to earth. From all of which it seemed obvious, to Eudoxos, that the earth was immobile – and, again, this was the most rational view in his intellectual context.

Further considerations which demonstrate the universal centrality of earth are these (all well known to Eudoxos): if the earth were not at the center of the celestial axis, it would necessarily be in but one hemisphere of the total celestial spheroid. Then, however, the nearest stars must appear brighter than those in the opposite hemisphere; were we nearer to Polaris then *Draco*, *Cepheus*, *Cassiopeia*, and *Camelopardalis* would appear immensely bright – much more conspicuous than *Virgo*, *Aquarius*, *Libra*, and *Orion*. And as one traveled from north to south, the polar stars would become faint, whilst equatorial stars would brighten. None of these things happen. The reasonable conclusion for an empirically-minded inquirer? Earth is immobile, in the exact center of a stellar sphere which turns uniformly on its axis once every twenty-four hours.

These natural, highly plausible hypotheses were beautifully embodied in the astronomy of Eudoxos, in accordance with the principle of perfectly circular planetary motion. The spheres of Eudoxos' sphere-systems were situated one inside the other, all concentric with the center of the earth. He placed each planet (or the sun or the moon), on the equator of a hypothetical sphere rotating uniformly about its own poles. The halts and loops of the planets, and their oblique movements across the ecliptic, were accounted for as follows: the poles of this hypothetical planetary sphere are fixed into a larger concentric sphere which rotates at some velocity other than that of the first sphere. And the poles of this second sphere were then set inside yet a third – again concentric with and larger than the first two; this third sphere had *its* own poles and speed of rotation. By a suitable choice of poles and angular velocities, Eudoxos was able thus to represent the apparent motions of the sun and the moon, with three spheres for each system. Let Aristotle present the *rationale* behind this choice of spheres:

Eudoxos supposed that the motion of the sun or of the moon involves, in either case, three spheres, of which the first is the sphere of the fixed stars, and the second moves in the circle which runs along the middle of the zodiac, and the third in the circle which is inclined across the breadth of the zodiac; but the circle in which the moon moves is inclined at a greater angle than that in which the sun moves. And the motion of the planet involves, in each case, four spheres, and of these also the first and second are the same as the first two mentioned above (for the sphere of the fixed stars is that which moves all the other spheres, and that which is placed beneath this and has its movement in the circle which bisects the zodiac is common to all), but the poles of the third sphere of each planet are in the circle which bisects the zodiac, and the motion of the fourth sphere is in the circle which is inclined at an angle to the equator of the third sphere, and the poles of the third sphere are different for each of the other planets, but those of Venus and Mercury are the same.[77]

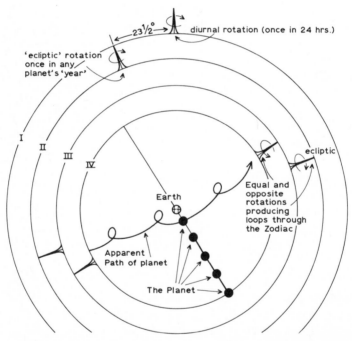

Fig. 14.   Eudoxos' Nester-Sphere construction; his attempt to simulate the annual motion of a typical planet (first approximation).

[77] Metaphysics, Λ1073b–1074a.

A typical arrangement for a planet was as in Figure 14. Such a quadrupartite configuration of computation-spheres was thought by Eudoxos to be independent of those required by any other celestial body. This is obvious, each requires its own computations for determining its elements. Eudoxos' scheme is not *systematic* at all, therefore, save only in that he employs but this one computational device for all celestial problems. He can deal with the motions of all planets, one at a time, but never altogether. This constitutes neither a cosmological system, nor even an astronomical system. The machinery imagined as *possibly* accounting for the motions of a planet permits no integrated cosmological picture of all celestial bodies at once, not in their composite spatial relationships to the earth. Nor are Eudoxos' calculations concerning any one planet dependent upon, or connected with, those concerning any other celestial body. The entire scheme is fundamentally just a calculating device for Eudoxos.[78]

It is not likely that Eudoxos was too successful with this scheme in predicting the positions of the planets. Thus Sosigenes remarks:

Nevertheless the theories of Eudoxos and his followers failed to save the phenomena, and not only those which were first notices at a later data, but even those which were before known and actually accepted by the authors themselves... I refer to the fact that the planets appear at times to be near to us and at times to have receded... For the star called after Aphrodite [i.e., the planet Venus] and also the star of Ares [Mars] seem, in the middle of their retrogradations, to be many times as large, so much so that the star of Aphrodite actually makes bodies cast shadows on moonless nights. The moon also, even in the perception of our eye, is clearly not always at the same distance from us, because it does not always seem to be the same size under the same conditions as to medium ... There is evidence for the truth of what I have stated in the observed facts with regard to total eclipses of the sun; for when the center of the sun, the center of the moon, and our eye happen to be in a straight line, what is seen is not always alike; but at one time the cone which comprehends the moon and its vertex at our eye comprehends the sun itself at the same time, and the sun even remains invisible to us for a certain time, while again at another time this is so far from being the case that a rim of a certain breadth on the outside edge is left visible all around it at the middle of the duration of these eclipses. [This is the first account ever given of an annular eclipse.] Hence we must conclude that the apparent difference in the size of the two bodies observed under the same atmospheric conditions is due to the inequality of their distances [at different times] ...[79]

If we disregard Sosigenes' acute criticism (and even some earlier remarks of Aristotle), Eudoxos must still be judged to have offered *some* intelligible account of certain anomalies which had been blocking any

---

[78] An argument to the contrary exists, but I cannot linger over it here.
[79] In Simplicius, Commentary on Aristotle's *De Caelo*, pp. 504–505 (Heiberg).

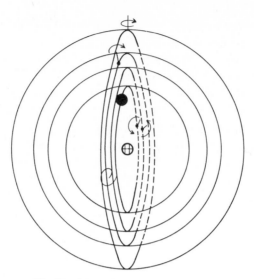

Fig. 15.   Mars: Ensphered *à la* Eudoxos.

"Here is how Eudoxos accounted for the motion of a planet, with four spheres. The planet itself is carried by the innermost, embedded at some place on the equator. The outermost of the four spins round a North-South axle once in 24 hours, to account for the planet's daily motion in common with the stars. The next inner sphere spins with its axle pivoted in the outermost sphere and tilted $23\frac{1}{2}°$ from the N–S direction, so that *its* equator is ecliptic path of the Sun and planets. This sphere revolves in the planet's own 'year' (the time the planet takes to travel round the zodiac), so its motion accounts for the planet's general motion through the star pattern. These two spheres are equivalent to two spheres of the simple system, the outermost sphere of stars that carried all the inner ones with it, and the planet's own sphere. The third and fourth spheres have equal and opposite spins about axes inclined at a small angle to each other. The third sphere has its axle pivoted in the zodiac of the second, and the fourth carries the planet itself embedded in the equator. Their motions combine to add the irregular motion of stopping and backing to make the planet follow a looped path. The complete picture of this three-dimensional motion is difficult to visualize." [80]

attempts to depict the universe as a set of spheres encircling the earth.

The Greek natural philosophers are often criticized for not making more observations of the heavens before drawing up their speculative, cosmical schemes. They would not have understood this criticism. There was already *more than enough observational data* before them, it all required explanation. To find some general intelligibility within the world

[80] E. Rogers, *Physics For The Inquiring Mind*, p. 229. Let us hope that the next ten pages lessen the 'difficulty to visualize' Eudoxos' scheme.

of their experience was their problem; this required reflection, not more data; the man entangled in taffy doesn't need more taffy. Why search for new difficulties when so much thought was still necessary to explain, and discover, 'the plan behind' natural events on all sides? The Greeks were born into a complex, data-jammed world requiring understanding. Most of Eudoxos' precedessors wished simply to find some intelligibility, if they could, in the one major fact of stellar and solar circular motion. They wished to see some structure behind these mighty events. It was their hope to become clear about what *kind* of thing was happening above them each day. They sought to show how celestial phenomena could be construed as a possible effect of processes familiar to them on a smaller scale. And, indeed, the northern heavens *do* seem somehow less puzzling, when the apparently random stars group together into familiar, named constellations which are then thought of as lights set into an enormous sphere turning around us. The night becomes somewhat less of a 'blooming, buzzing confusion' when so ordered, named and 'explained'. The Greeks could not do everything at once; they felt a need to explain what was, and had been, happening all around them; thus before they could set out to predict what had not yet happened – and before they could set off in search of new *explicanda*. They had to perceive the 'plan' in nature before they could try to gain power over it.

By Eudoxos' time not only the stellar motions and the 'interdependence' of sun and moon, but also perplexities about the looping reversals of the planets had to be reckoned with. Eudoxos suggests a geometrical arrangement of spheres which, were they but imagined to exist in the heavens, could give something like the same sort of resultant motions. He is trying only to understand what *kind* of behavior these stellar gyrations *might* be. His immediate concern is not to foretell future examples of it.[81] Eudoxos sought some familiar arrangement of familiar objects which could generate resultant motions rather like those observed for Mars, and for Venus. In this he achieved a measure of success.

Thus the *sun* was supposed to have a motion analogous to that of a point on the inner member of a three-sphere system. The outer one (1) had the daily motion of the fixed stars. The second one (2) turned slowly along the zodiac. The third (3) carried the sun (at its equator), and moved

[81] To inquire what kind of bird just flew over, is obviously not to ask where others like it might be found.

it along a great circle *slightly inclined* to the zodiac at the rate of one turn a year.[82]

The outer sphere (1) in the fourfold system used for each *planet*, gave the body's motion around the earth in twenty-four hours. (Cf. Figure 14

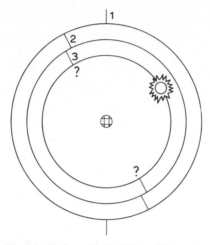

Fig. 16.   Eudoxos' Solar Sphere-Cluster.

and 15.) The second sphere (2) corresponded to its uniform motion along the zodiac.[83] For Mars, Jupiter, and Saturn, the turning rate of the second

---

[82] Eudoxos actually rejected the discovery of the variable orbital velocity of the sun made by Meton and Euktemon seventy years earlier. Instead, he made the sun travel not along the ecliptic, but along a circle inclined to the latter. According to Simplicius (p. 493, L 15 (Heiberg)):

> "Eudoxos, and those before him, were led to make this assumption by observing that the sun at the summer and winter solstices did not always rise at the same point on the horizon."

But this 'observation' is certainly the result of inaccurate naked-eye determinations of the azimuth of the rising sun, and crude sightings with the rude gnomon. Hipparchus actually quotes Eudoxos as saying:

> "It seems that the sun also makes its return in different places, but much less conspicuously." (Enoptron).

[83] Eudoxos had no knowledge of the orbital changes of velocity of planets which depend, of course, on the eccentricity and ellipticity of each orbit. Nor did he take the orbits to be at all inclined to the ecliptic; he simply let the second sphere of each planet move along this circle. Latitudes of the planets were supposed to depend not on their longitude, but only on their elongation from the sun. It was in order to represent this north-south motion in latitude, and also the inequality in longitude depending on elongation, that the third and fourth spheres were originally introduced.

sphere equalled the time the planet required to circle the heavens from some arbitrarily designated position (against the fixed stars), back again to that same position. For Mercury and Venus – 'the sun-linked ones' – the rate of turn was just one revolution per year, as with the sun itself. The axis of the third sphere was fixed into the equator of the second. Its rate of turning equalled the interval between two successive oppositions or conjunctions with the sun. (Cf. Figure 17 and 18.)

The axis of the fourth sphere was inclined slightly to that of the third, at an angle specially suited to each planet and its average deviation from the ecliptic. It turned at the same rate as the third sphere, but in the opposite direction. On the equator of this fourth sphere the planet itself was fixed. As a result of this composition of motions a planet would then show (1) a daily westward movement around the earth (the result of the first 'outer' sphere), (2) a 'yearly' eastward motion along the zodiac (from the second sphere), and (3) two other motions (from the third and fourth

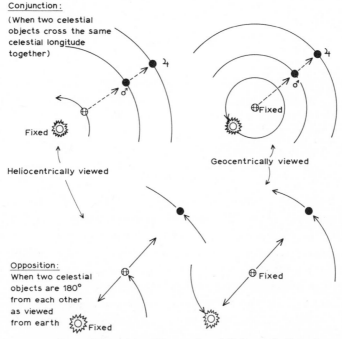

Fig. 17.   Conjunction and Opposition: Geocentric and Heliocentric. Some terms needed in understanding Eudoxos' and all subsequent planetary theory.

SUPERIOR AND INFERIOR CONJUNCTION OF VENUS AND THE SUN

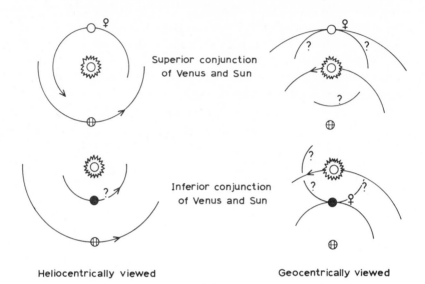

Superior conjunction
of Venus and Sun

Inferior conjunction
of Venus and Sun

Heliocentrically viewed                    Geocentrically viewed

Fig. 18.    Eudoxos' spherical Lemniscate, or 'Hippopede'.

spheres) occurring between successive oppositions and conjunctions.
The motion imparted by spheres 3 and 4 conjointly is, as has been demonstrated conclusively by Schiaparelli, that of a spherical lemniscate
(cf. Figure 18); the contour of this geometrical configuration is best
imagined as the surface outline of a smooth hole drilled through the very
edge of a solid sphere.

The Cartesian equation for this figure (a figure of eight) considered
in a *plane* is $r^2 = a^2 \cos 2\theta$. But, we must now imagine this figure to be
projected on the (outer) surface of a sphere. The curve's longitudinal
axis is imagined to lie along the zodiac; its total length equals the diameter of the circle described by the pole of the 'wobbling' sphere (4)
which actually 'carries' the planet. The point of crossing is 90° from the
two poles of rotation of the third sphere. The planet describes this curve
by moving over equal arcs in exactly equal times.[84]

Now think of a planet in motion along this curve generated by spheres
3 and 4. Combine with this the further motion given to it by the second

[84] This method of constructing Eudoxos' 'hippopede' as he called it, is given in Simplicius,
*op. cit.*, p. 497.3; cf. also Proklus, *Commentary on Euclid's Elements*, Bk. I, p. 112.5.

sphere. This would be somewhat like fixing the solid sphere illustrated in Figure 19 (with its 'drilled' lemniscate outermost), onto the spinning circumference of an immense roulette-wheel. As the planet traverses the spherical lemniscate, and as the roulette-wheel pushes the lemniscate itself through a wide, fast circle, the resultant path of the planet through space will correspond to its apparent motion through the zodiacal constellations. Imagine the planet to be a brightly glowing light-source

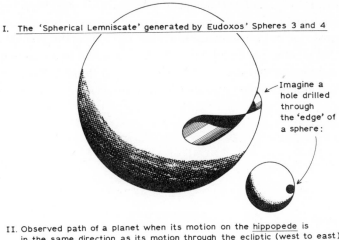

I. The 'Spherical Lemniscate' generated by Eudoxos' Spheres 3 and 4

Imagine a
hole drilled
through
the 'edge' of
a sphere:

II. Observed path of a planet when its motion on the hippopede is
in the same direction as its motion through the ecliptic (west to east)

ecliptic motion

III. ... when its motion on the hippopede is in the direction opposite to its
motion through the ecliptic

Fig. 19.   Mechanism behind Eudoxos' 'Trace-in-Space'.

in an enormous dark room; the linear trace thus perceived could, by tinkering with the geometry and dynamics of this roulette-and-sphere machinery, be made to approximate to *any* planet's actual trace in the heavens. (Cf. Figure 19, bottom, and Figure 20.) When the planet moves backwards on the lemniscate more rapidly than the latter is being moved forward on the roulette-wheel, then the planet will appear to an observer (at the system's central hub) to have a retrograde motion. When the two opposite motions are momentarily equal, the planet appears to be stationary. Finally it goes forward again. And, of course, this entire roulette-and-sphere machine is itself fixed within a still larger spinning disc – like a railway's roundhouse turntable.

This turntable, then, adds (1) the *diurnal* twist to (2) the *annual* motion imparted to the planet by the roulette wheel, and to (3) + (4) the *zodiacal* motion up, down, forward and backward imparted to it along the spherical lemniscate. Clearly, the greatest forward accelerations will occur at the 'cross-over' of the lemniscate ($\alpha$ in Figure 21) and at this point also will occur the greatest retrogradations ($\beta$). Eudoxos had to adjust and time these motions so that the planet passes forward through this point just when it is in superior conjunction with the sun (on the sun's far side along

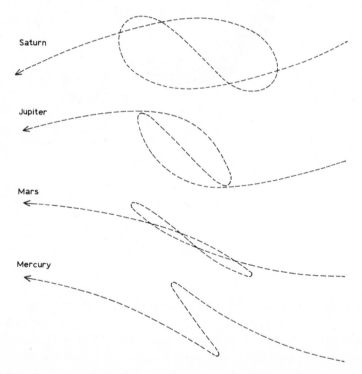

Fig. 20. Eudoxos' paths for four planets: exactly calculated from original sphere ratios (after G. V. Schiaparelli, in *Geschichte der Mathematick*, Leipzig 1877, Plate II).

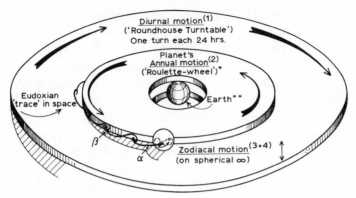

*E.g. Saturn's Revolution takes 29¹/2 years

**The immobile observation platform

Fig. 21.   Eudoxos' constructions for Saturn, Jupiter, Mars, and Mercury.

the line earth-sun-planet). It must also pass backward through the point when the planet is at inferior conjunction (between sun and earth, along the line earth-planet-sun). These motions are naturally accompanied by an up-down trans-zodiacal movement which will vary as does the depth of the lemniscate. (Cf. Figure 20.)[85]

Eudoxos did not seek the physical causes of these rotations. Nor does

[85] The hippopede (ιππουπεδη), or 'horse fetter' was so-called because of the similarity of its '8'-shaped geometry with that of a trotting exercise familiar in Athenian cavalry

Proklus' conception of the 'Hippopede'.

academies. The word occurs also in Proklus' *Commentary to the First Book of Euclid*; there he describes the plane sections of the solid generated by revolving a circle around

he seem to have thought that such sphere-nests really existed in three-dimensional space; *a fortiori* he is not concerned with what material such spheres might be made of, and how far from each other they are situated. Moreover, Eudoxos never relates the motions of one set of spheres, e.g., Mars', with those of any other system, e.g., Venus. This would have been difficult to do; the first sphere of each system (which gives the diurnal motion), must either be identified with the actual stellar sphere – so that it would be common to all individual sphere-systems *or* each local system must have its outer sphere mechanically co-axial with the corresponding outer sphere of every other local system. Neither possibility affords any physically picturable state of affairs: this fact complicates Aristotle's Cosmology.[86]

No; Eudoxos' sphere-clusters are geometrical formulae only – mere computational devices. They serve him *not* as some possibly real cosmic machinery, but only as an aid to the imagination – an aid which reveals the jogging motions of the planets as being a composite effect of familiar, intelligible motions. In Eudoxos, the planets' crazy loopings cease being wholly recalcitrant, planless, baffling anomalies. Being understandable as results of trains of perfectly circular motions, there was no need to fancy these paths as the whim of divine caprice, or some randomness in the cosmos. The principles of planetary motion need be fundamentally no different, then, from those of the obvious stellar motions. To have shown this much was, for Eudoxos, to have supplied a supremely rational solution to Plato's problem.[87] He demonstrated conclusively (what needed no demonstration for Greeks) that nature is profoundly ordered and intelligible structured, whatever its casual appearances may suggest.

In two ways, therefore, Eudoxos differs from his predecessors. While they merely speculated that there *must be* some intelligible structure to the heavens, he showed in geometrical detail what that structure *might* be like. And while they sought a single picture to reveal the complete framework of the cosmos in one stroke of imagination, he felt this to be

---

a straight line in its plane, assuming the line to be tangent to, but not cutting, the circle. Thus this major section of the 'anchor ring' is called by *hippopede* by Proklus. Hence, Eudoxos and his associates knew well the properties of this curve resulting from the combined motion of the third and fourth planetary spheres. (Compare the account of Theon of Smyrna, p. 328 of the Martin edition of his works.)

[86] Cf. pp. 62–84.

[87] Cf. Simplicius, *De Caelo*, p. 488 (Heiberg).

impossible. Eudoxos is the first theorist who copes with but one anomaly at a time. That the several resultant explanations do not fall into one pattern was less important to him than suggesting how each apparently irrational event could be reduced to something orderly and intelligible: in short, something possible. This is reminiscent of Dray's analysis of some explanations: α is explained when some *possible* account of how it came about has been generated when α's occurrence ceases being a thorn in the understanding.[88] Even the gods could not contradict themselves, so even a completely hypothetical resolution of an apparent contradiction seemed a pale reflection of cosmic intelligibility.

For Saturn, Jupiter, and Mercury, Eudoxos' system accounted well for motions in longitude. It was rather unsatisfactory for Venus; with Mars it broke down altogether. The limits of motion in latitude (up↔down) were well represented by the several hippopedes, although their periods and places in the cycle were rarely correct. But that Eudoxos could not predict planetary behavior with accuracy remained secondary. What did matter? His geometrical inventions revealed that *in principle* the planets' wanderings could be accommodated within some intelligible scheme of the universe – which Greek reason required to be geocentric, geostatic and constituted of circular motions. He had fulfilled the Greek ideal: that's what mattered. The world was again thinkable. Moreover, Eudoxos' entire system appeals to but three independent elements: (1) the epoch of a superior conjunction, (2) the period of sidereal revolution, and (3) the inclination of the axis of the third sphere to that of the fourth. With these alone he built his astronomical 'computing machines'. For the same task today we require six elements, not three.

Eudoxos' associate, Kalippus, made more exact the correspondence between geometrical models and phenomena observed. Kalippus felt that more spheres than Eudoxos used were now unavoidably necessary.

Kalippus made the position of the spheres the same as Eudoxos did, but while he assigned the same number as Eudoxos to Jupiter and to Saturn, he thought two more spheres should be added to the sun and two to the moon, if one seeks to explain the observed facts[89], and one more to each of the other planets.[90]

Kalippus added one sphere to each Eudoxian system for Mars, Venus,

[88] Cf. p. 14, *supra*.

[89] This probably refers to the inequality of the seasons; see Eudemus, in Simplicius' *Commentary on Aristotle's De Caelo*, p. 497.17.

[90] Aristotle, Metaphysics Λ.

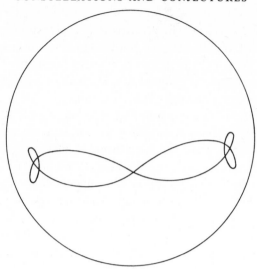

Fig. 22.   Kalippus' complicated Lemniscate.

and Mercury. The purpose of these was to supply a retrograde motion without ruining the synodic motion, as Eudoxos' third and fourth spheres by themselves had done.[91]

This provided planetary paths which traced not a simple spherical lemniscate but a curve of an intricate contour (see Figure 22).

When 'stretched' around the zodiac (by the second sphere), this curve met *some* discrepancies within Eudoxos' scheme. To account for the sun's unequal motion in longitude, Kalippus introduces two new spheres functioning much as did III and IV for the planets. This accommodates the Fifth Century (B.C.) discovery of the inequality of the seasons. Or, as Geminus put the point later:

Why, although the four parts of the zodiac circle are equal, the sun, traveling at uniform speed all the time, yet traverses the arcs in unequal times.[92]

Similarly, Kalippus' lunar theory gains two additional spheres, presumably to embrace the elliptic inequality. Evection not yet having been discovered, Kalippus' theory was thus as good as any other. His additions reconciled Eudoxos' geometrical hypotheses with the observed positions of sub-stellar celestial bodies (well within the limits of accuracy then attainable).

[91] The *synodic period* for the inner planets is the span from one inferior conjunction with the sun to the next: for the outer planets it is the time between two oppositions to the sun.
[92] *Elements of Astronomy*, I. (trans. T. L. Heath).

*Aristotle*

> Aristotle's generalization of Eudoxos' system
> Fifty-five or sixty-six spheres?

*Aristotle as a Cosmologist*

> Aristotle as astronomer
> Aristotle as physicist
> Aristotle as mathematician

*The Writings of Geminus*

Aristotle (384–322 B.C.) not only accepted that the universe was geocentric, geostatic, and fundamentally circular, he *argued* for these things with an ingenuity and thoroughness never before encountered. Indeed, seldom since. Aristotle begins by rejecting the Pythagorean supposition that the most excellent body must occupy the center of the universe.

For that which is to be defined is the middle, but what defines is the extremity; and more excellent is that which encompasses and is the limit, than that which is completed: the latter is matter, the former is the essence of the composition.[93]

Aristotle then rejects the idea of Thales, Anaximenes, Anaxagoras, and Demokritus – that the earth's stability *follows* from supposing that it floats on water, or that it is flat. Similarly, Empedokles' notion that the earth moved to the center of a whirling universe is also untenable: for why should everything heavy tend toward the earth when the whirling takes place so far from us? And why does fire tend upwards, away from us? The condition and position of the earth cannot be a consequence simply of the motion of the heavens. Anaximander had suggested that the earth could not fall in any one direction because it is placed in the middle – and has the same relation to every part of the circumference. But this is not enough, muses Aristotle. Because *by its nature* earth moves toward the center. It is not its relation to the circumference that keeps earth at the center of the universe; rather, because every part of the earth is *of its essence* such that if lifted from its surface it will fall again towards the earth's center – *there's* the reason for earth's centrality.

Does the Earth itself move?

... some make it out to be one of the stars, while others place it in the middle and assert that it is packed and moved around the middle axis.[94]

But such a motion cannot naturally belong to earth, since then it would also be natural for the earth's individual parts. (Another singular argu-

[93] *De Caelo*, II, 13, p. 293b.
[94] *De Caelo*, II, 14, p. 296a.

ment.) Instead, we see individual parts of the earth move in straight lines towards its center.

Furthermore it appears that the bodies which have circular motions in space, all, with the exception of the first sphere, also have a backward motion, and in fact have more motions than one, so that necessarily also the earth, whether it moved around the center [as in Philolaus' theory] or was moved at the center itself, would move in two courses; but in this case there would of necessity be passings by and turnings of the fixed stars; but none such appear; the same stars both rise and set at the same places.

Here Aristotle seems unaware that anyone had ever accounted for the daily motion of sun, moon, and stars by assuming a rotation in the earth itself. For him, the center of the earth coincides with that of the universe; heavy bodies do not fall in parallel lines, but along paths converging towards one center – that of the earth. Bodies projected upward, fall straight down on the point from which they started. Hence earth is neither in motion nor resides anywhere but at the center. And since its parts are by nature meant to move from all sides toward the center, it is not possible that any part of earth could be moved permanently away from the center; therefore the whole earth cannot do so either. And as clinching proof of the earth's immobility Aristotle notes that all phenomena are in fact observed exactly as they ought to be observed on the theoretical supposition that the earth really is at the center.

Aristotle argues for the sphericity of the earth, as his predecessors had done; he observes that when objects are moved uniformly from all peripheral points towards a common center, a single aggregate body forms whose surface is everywhere equidistant from this center. Even were the particles unequally propelled inwards, the larger ones would push the smaller ones on ahead, until the whole was everywhere uniformly and compactly settled around this center.[95] He notes also our circular shadow in lunar eclipses; our earth must therefore be a sphere. Moreover, even a terrestrial journey north or south will change the horizon sensibly and will place different stars overhead; 'Egyptian stars' are not seen farther north, while northern circumpolar stars in the north rise and set in the south.[96]

Aristotle was no mathematician. Nor was his outlook even affected by mathematical considerations, as Plato's surely had been.

In his geometrical computations Eudoxos had entertained imaginary

---

[95] *De Caelo*, II, 14, pp. 297b–298a.
[96] Compare Newton's 17th century adaptation of the same arguments; cf. p. 46 *supra*.

spheres solely to 'account for' the observations; to 'reduce' the apparent anomalies to possible intelligible structures.[97]

Although he was aware of Eudoxos' intentions, Aristotle could not rest with suggestions so tentative and hypothetical. For him the spheres actually existed. For the Philosopher the spheres' very intelligibility guaranteed their existence.[98]

Aristotle's spheres were three-dimensional crystalline shells, parts of the physical machinery which keep celestial bodies in motion. Eudoxos and Kalippus may have satisfied themselves with purely formal analogues of the universe. Their ideas of 'the intelligible' were fulfilled by fixing the errant planets into an abstract geometrical framework. Aristotle would not settle for this. Kalippus was relieved when his additional spheres accounted for what Eudoxos' third and fourth spheres had failed to explain. Aristotle could not be calmed by any such patchwork, so formal and arbitrary. He refused to identify as descriptive physics the speculative hypotheses of mathematicians. Nothing short of actually linking the motions of all the spheres – of forming a single, comprehensive, cosmological machinery – could meet Aristotle's requirements. If Eudoxos' partial plottings 'saved the phenomena' – i.e., made them thinkable – then why not piece them all together into one single master-plot? Aristotle wanted to construct a composite celestial mosaic, of which Eudoxos and Kalippus had formed only the isolated and unrelated tiles. He wanted a universe, not a pluriverse! Intelligibility, for Aristotle, could not consist *merely* in resolving the perceptual complexities and conceptual perplexities of planetary behaviour by independent, *als ob* mathematical hypotheses – one at a time; as if astronomy were a plumber's jumbled toolbox of possible 'leakpluggers' Ultimately, Eudoxos'

[97] Here again is Dray's alternative to Hempel: one sometimes 'explains' a puzzling welter of phenomena by showing that they could be *possible* effects of some hypothetical, but wholly rational, imaginative configuration. This is not the same as inferentially relating an anomaly, back through laws, to initial conditions. It is more like perceiving a potential pattern in perplexing phenomena, even when that pattern has no initial claim to *de facto* existence. That *some* pattern is possible at least shows the data to be intelligible-in-principle, i.e. explicable ultimately. In Dray's sense α *is explained* when it is shown to be at least explicable. In Hempel's sense α *is explained* only when, *after* having been shown to be explicable, it is deductively linked to initial conditions. In Dray's sense Eudoxos, and Aristotle, were certainly seeking to explain the celestial phenomena.

[98] For us, on the other hand, existence usually determines intelligibility – by the principle 'P exists therefore P is possible'.

solutions (if tenable at all) had to follow systematically from philosophical principles about the nature of the universe.

Aristotle pondered well the systems of Eudoxos and Kalippus. How to prevent the motions of the 'outer' sphere-systems from mechanically overriding the 'inner' sphere-systems? There could not be one single, solid axle giving each system its characteristic diurnal motion – not unless the individual sphere-systems were strung out like chops on a spit. How could one get *all* the diversely-rotating spheres physically concentric with the earth, and yet keep them out of each other's way – all impaled by one solid celestial axis. And how to adapt all of this to Aristotle's major principle, whereby the motive power works from the periphery of the universe towards its center?[99]

Aristotle's ingenious suggestion was this: that between the innermost sphere (D) of each planet-system (the planet-carrier) and the outermost sphere of the next inferior planet (α: see Figure 23), there must be a set of spheres whose counter-rotations, or 'unrollings' neutralized the effect of the superior planet's motion. Otherwise this latter would override the next inferior cluster. The relevant passage is in Aristotle's *Metaphysics*.[100]

... If all the spheres [of Eudoxos and Kalippus] combined are to explain the observed facts, then for each of the planets there should be other spheres (one fewer than those hitherto assigned), which counteract those already mentioned and bring back to the same position the outermost sphere of the star which in each case is situated below the star in question; for only thus can all the forces at work produce the observed motion of the spheres. Since, then, the spheres involved in the movement of the planets themselves are eight for Saturn and Jupiter and twenty-five for the others, and of these only those involved in the movement of the lowest-situated planet (the moon) need not be counteracted, the spheres which counteract those of the outermost two planets will be six in number, and the spheres which counteract those of the next four planets will be sixteen: therefore, the number of all the spheres – both those which move the planets and those which counteract these – will be fifty-five ...

The Figures 23 and 24 should make this passage crystalline-clear.

Spheres A, B, C, and D are as I, II, III, IV, depicted earlier in Eudoxos' scheme. But sphere D′ is meant to 'unroll' sphere D; it turns on the same

---

[99] The argument is as elucidated by Simplicius (*De Caelo* II, 12, p. 497 (Heib.)); Sosigenes is given there as the authority. Note again the reasoning: since the circumference of a circle determines the properties of its center, and since the motive power of the universe works from the periphery towards its center, therefore the earth must have the same shape as the cosmos. The fixed stars show us that the cosmos is spherical; so the earth must be spherical.

[100] Λ1073b17–1074a14.

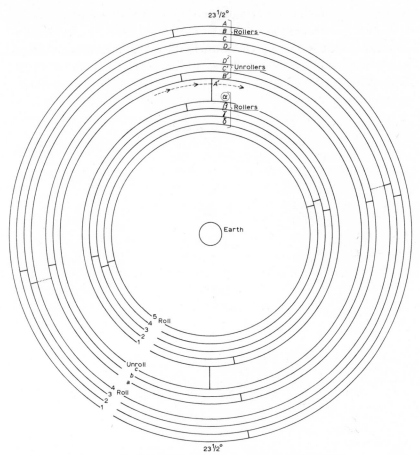

Fig. 23.   A Eudoxian Sphere-Cluster adapted by Aristotle.

geometrical axis as D, but in the opposite direction. So any point on D′ might as well have been attached directly to C. But then sphere C′ turns on C's axis; again in the opposite direction. Thus any point on C′ could have been on B. Now sphere B′ turns against sphere B on the same axis again. So any point on B′ could have been on A itself, which turns with the fixed stars. The air is therefore clear for the next inferior sphere-system, let us say that of Mars, ♂.

But why, then, set Mars' sphere system inside B′ (which now turns with the fixed stars)? Mars' outer sphere *also* turns with the fixed stars.

Is not this redundant? A lengthy historical tradition answers 'Yes'.

Particular attention should thus be given to this question – and to the spheres labelled B′ and α. The former (B′) is the innermost 'unroller' (ἀνελίττουσαι) of e.g., Jupiter's system; the latter (α) is then the outermost 'roller' of Mars' system, and gives it its diurnal revolution. The point of the argument to follow is honed between these two spheres. Do

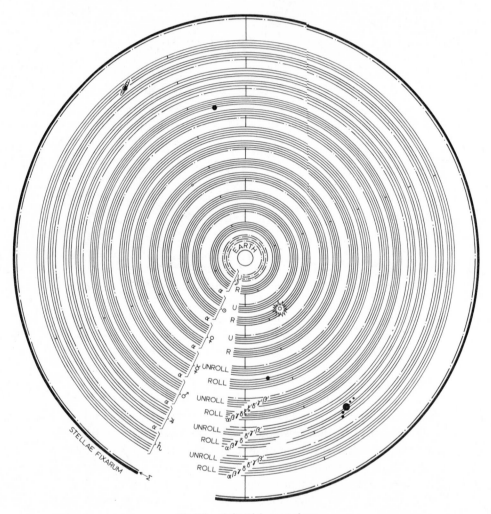

Fig. 24.   Aristotle's cosmology.

they, in Aristotle's terms, do the same work or too much work, or too little?

In *De Caelo*[101] all these spheres are said to be in physical contact. But the above representation is clearer for our purposes.

For 2300 years Aristotle's cosmology has been described (beginning with his own *Metaphysics*) as comprising 55 astral spheres. Commentators have remarked in this number six redundancies, namely, those spheres (α) which give to each sphere-cluster (below Saturn) a diurnal rotation. But it has been supposed universally that Aristotle's 55 spheres *can* really generate the celestial motions he requires; his achievement is usually granted, but it is merely described as somewhat 'inelegant', 'uneconomical', 'superfluous', or 'redundant'.

In fact, the idea that 55 spheres can achieve what Aristotle wants, even 'redundantly', is demonstrably false. The required observations can be generated either within a system of 49 spheres, or within one of 61 spheres, the latter being preferable. A 55 sphere system simply cannot work. The dotted lines in Figure 24 represent spheres not mentioned by Aristotle. One of our objectives now will be to show that these extra spheres are necessary to fulfill The Philosopher's original aspirations. For the moment, however, let us fix our attention again on the 55 spheres as Aristotle actually describes them.

Concerning this number Dreyer writes:

Obviously he [Aristotle] might have practiced economy a little by leaving out six spheres...[102]

Heath writes:

Aristotle could therefore have dispensed with the second of these, namely the first of the spheres belonging to the inner planet, without detriment to the working of his system ...[103]

Sir David Ross writes:

In this system, however, both the innermost reacting sphere of a planet and the next sphere to it, the outermost deferent sphere of the next planet, are moving with the same motion, viz. that of the fixed stars, so that the second of these two spheres is superfluous. Aristotle might thus have reduced the total number of spheres by six ...[104]

Obviously Dreyer, Heath and Ross all feel that a fifty-five sphere system

[101] 287a, 5–11.
[102] J. L. E. Dreyer, *A History of Astronomy* (Dover) p. 113. That is, Dreyer is concerned with the 'redundancy' of Jupiter's B′ and Mars' α as pointed out in our earlier exposition.
[103] T. L. Heath, *Aristarchus of Samos* (Oxford, 1913), p. 219.
[104] W. D. Ross, *Aristotle's Metaphysics* (Oxford, 1924), Commentary, p. 392.

would work, but would be inelegant. We will now show that, elegantly
or otherwise, a 55 sphere cosmos cannot even work – not as Aristotle
describes the motions and dispositions of the constituent spheres.

Consider the 55 spheres as Aristotle describes them. As we said, the
sphere α in Figure 23 gives the diurnal rotation to each quadrupartite
sphere-cluster. In Eudoxos' representation this sphere turned around its
own poles in absolute space. In Aristotle's system there are but two possibilities for α; either it does rotate about its poles, or it has no motion
at all. Aristotle is explicit in adapting Eudoxos' technique *en bloc* for
his own cosmology. Hence, it seems reasonable to study the first alternative, viz. that α rotates on its own axis.

As Aristotle says:

"... The first is the sphere of the fixed stars ..." [105] by which he "... means not that the first
sphere of the sun or of the moon was the sphere of the fixed stars, but that it had the same
motion ..." [106].

So our first assumption will be that α turns on its own 'axle' at the same
rate as the sphere of the fixed stars turns around its poles. This means that,
no matter where or how α's poles may be fixed into some other sphere, α
will always turn on its axle once in 24 hours, whatever the axle itself may
be doing. It is as if we had a snail crawling around a wagon's axle once
every 24 hours, whether or not the wagon was motionless. If the wagon
axle itself turns, and the snail's speed remains constant, then the snail's
*absolute* speed of rotation will be its own (1 turn/24 hours) *plus* the speed
(? turns/24 hours) of the axle itself.

In fact, *every* α, i.e., the outermost sphere of every sphere-system,
is fixed into some other sphere. With all the heavenly bodies, save Saturn,
α is fixed within a sphere B'. What, again, is the motion of B'? It 'unrolls'
the original motion of sphere B – the second in any planetary system.
Like B, its poles are fixed 23.5° of arc from those of the diurnally-rotating
outer sphere. Thus, if (by reverse-turning) B' cancels out the motions of
B, then any point fixed within B' will move as if it were a point in A itself
(Jupiter's outer sphere in Figure 23). So all points constituting sphere B'
will rotate on, or revolve around, the axle of B' once every 24 hours.
Now α (Mars' outer sphere in Figure 23) is fixed within B at 2 points.
These points must, therefore, rotate or revolve once every 24 hours – by

[105]   1073b19.
[106]   Ross, *op. cit.*, p. 387.

the considerations just advanced. By doing so, the axle of α will also spin (because the points at which it is fixed in B' do so) once every 24 hours.

Now, if α were absolutely *motionless* on its axle, then the fact that its axle is fixed in B' (which moves as do the fixed stars) means that any point on α will have the diurnal rotation along with B'. However, α is nowhere described by Aristotle as being motionless on its axis. Rather, it rotates on its axle once each 24 hours. It does this irrespective of what further motions are communicated to that axle by superior spheres. This means that any point on α will have not just a diurnal revolution around α's axle, but also a further increment of angular velocity resulting from the *axle itself* (or, better, the axle-bushings which are part of B') having been given a diurnal rotation by B'; just as in our earlier snail-on-axle example where the snail's (α) absolute motion had to be reckoned as the sum of its own local motion around the wagon's axle (B'), and the additional motions of the axle itself.

If α is the same type of sphere as all the others in the system, it will have two motions: (1) a local motion about its own axle, and (2) the axle itself will be moved by its next superior sphere. Thus sphere B turns from west to east on its own axle at an angular velocity corresponding to the solar year of the planet in question. But meanwhile the axle of B, being set into A, is twisted from east to west diurnally. The kinematical composition of these two local motions generates a major part of the planet's projected motions in absolute space.

If α is in this way like the other spheres, then we must calculate both its local motion around its axle *and* the motion the axle itself is given by B'. But if α is unlike the others, we are nowhere told this by Aristotle, or anyone else; why should this be so, and in what way? We are so bemused by α and B' having the same *local* motions that we have for millennia neglected to add these local motions – as we do with all the other spheres. The resultant motion is *not* a diurnal rotation. Indeed, the idea of a motionless (or non-existent) sphere α – the idea behind any suggested 49 sphere cosmos which drops all six α's – is distinctly un-Aristotelian, as we shall see.[107]

---

[107] Cf. Ross:

> "Every perfect substance produces a motion in the heavens. Therefore the number of perfect substances is the number of the motions required to explain the motion of the heavenly bodies." (pp. 394–5).

In other words, between each quadrupartite planetary sphere-system, the sphere corresponding to B′ fails to accomplish what Aristotle required, viz. "... to neutralize (the positive spheres') action on the outer sphere of the next system (counting inwards)".[108] This neutralization could have been accomplished *only* by one more unrolling sphere, A′, for each system. (Cf. the broken lines in Figure 23.) Since A′ would turn from west to east once a day, it would cancel A's motion. Any point in sphere A′ would then be *at rest* in absolute space. Hence, had α's axle been fixed anywhere in an A′, *no* residual superior motion would have been conveyed to it. *This* was Aristotle's intention! Now α could indeed have the diurnal rotation, just as Aristotle describes; and the moon's motions could have been calculated closely to correspond with what we actually observe.

As things stand, however, the 55 sphere system set out in the *Metaphysics* has these properties: the innermost unrolling sphere of Saturn's system ( ♄ B′) itself has the diurnal rotation, as we have seen. The outermost rolling sphere of Jupiter ( ♃ α) also has a diurnal rotation on its own axle. When the latter is imbedded in the former, Jupiter's entire system must turn with *twice* the diurnal rotation. This must be so if Jupiter's outer sphere ( ♃ α) moves at all. But had he wished ( ♃ α) to be motionless, Aristotle would surely have described it thus.

Moreover, the innermost unrolling sphere of Jupiter ( ♃ β′ – corresponding to B′ in Figure 23), since it neutralizes the second rolling sphere of Jupiter ( ♃ β′), will be such that any point on it might as well have been on Jupiter's outermost sphere ( ♃ α). But we have seen already that the latter (in relation to Saturn) moves in 'absolute space' with *twice* the diurnal rotation; hence this must also obtain for the innermost sphere ( ♃ α) of this system. So Mars' outermost sphere ( ♂ α) is set within a sphere (i.e., ♃ α), which is already spinning with twice the diurnal rotation. But ♂ α is itself described as turning on its own axis once every 24 hours! The cumulative kinematical effect of all this will be for Mars' system to spin around the earth not once a day (as with Saturn and the fixed stars), nor twice a day (as Jupiter would do) – but three times each day!

The process is clearly cumulative: Mercury ( ☿ ) will circle Earth

---

[108] This is Ross' paraphrase of 1074a8.

four times a day, Venus ( ♀ ) five times a day, the sun ( ☉ ) six times a day – and the moon ( ☽ ) seven times each day! The 55 sphere cosmos as described by Aristotle, therefore, entails the moon's circling the Earth once every three hours and 24 minutes.

This is a conclusion Aristotle would certainly have wished to avoid. How? We hinted at one possibility: the 6 spheres α could be made motionless and rigid on their axles. If the outer spheres ♃α, ♂α, ☿α, ♀α, ☉α, and ☽α are to have any motion in absolute space, Aristotle wants it to be that of the fixed stars. But this can be achieved now only by denying to all these α's *any* local motion around their own axes. For, as we saw, if the α's both have a local motion of their own, and are set within spheres (β′) having *their* own local motion, the effect will be an absolute motion which increases below Saturn; until the moon ends up whirling around Earth every 204 minutes. In Figure 23, the diurnal motion of A is *not* neutralized by B′ but only reinstated. α adds to that motion another twist each 24 hours – and β′ does nothing to neutralize that. Each new α adds another spin – unresolved by any new β′. And so on, until the moon described in Aristotle's *Metaphysics* seems more like a Cape Kennedy production than like a familiar piece of the cosmos.

Even Heath confuses local motion and absolute motion:

But the first sphere of the next nearer planet (as of all the planets) is also a sphere *with the same motion as that of the sphere of the fixed stars*, and consequently we have two spheres, one just inside the other, with one and the same motion, that is doing the work of one sphere only.[109]

If the italicized expression refers to the first sphere's local motion about its own axis – and there is no reason to suppose it does not – then Heath's conclusion simply does not follow. The inner sphere will have *twice* the angular velocity of each β′ in absolute space. But if the italics refer to each α's absolute motion about its axis, there can be *no* mechanical, or kinematical, connection whatever between the spheres above β′ and those below – a conclusion wholly incompatible with Aristotle's broader objectives.

Compare Heath himself, writing in another place:

Aristotle ... transformed the purely abstract and geometrical theory [of Eudoxos and Kalippus] into a mechanical system of spheres ... this made it almost necessary, instead

---

[109] *Op. cit.*, p. 219; my italics.

of assuming separate sets of spheres, one set for each planet, to make all the sets part of one continuous system of spheres.[110]

Compare Ross:

Aristotle aims at a mechanical account and cannot isolate the system of one planet from that of the next.[111]

But giving any sphere α an *absolute* diurnal rotation must isolate it completely from sphere β', above it.

However, if α is fixed rigidly within β', the two might as well have been but one sphere to begin with. A sphere totally lacking local motion around its own axle would have no cosmological function in Aristotle's scheme. This point is established in a long tradition of Commentaries: it has repeatedly been suggested that 49 spheres would have served Aristotle just as well as 55. But let us be clear about this; 55 spheres, as Aristotle describes them, could not serve his purposes *at all*: the suggestion that β' and α be 'amalgamated into one' constitutes not just the tidying up of a trivially redundant system, but rather a fundamental overhaul without which Aristotle's scheme is immediately refuted by a few glances at the heavens. So (in Figure 23) we might try to drop α out of the scheme altogether, fixing β directly into B'. We could not, however, even begin to adopt the alternative of fixing α directly into C' (i.e., by dropping B' rather than α). The reason? The particular local motion of B' is indispensable for giving any point on that sphere an absolute motion comparable to what it would have been if set directly into the sphere of the fixed stars. C''s local motion is obviously different from that of B'. So fixing α into C' would result in an absolute motion for α wholly different from what is required, namely, a combination of the next superior planet's 'cancelled' solar year plus a diurnal rotation. ♂α set into ♃γ' would give, instead, a diurnal rotation ( ♂α's) plus Jupiter's inclination to the ecliptic – plus the motion of ♃β, as yet uncancelled. ♂α's kinematics would be ruined by this. So the 'sphere-dropping' move *must* consist in fitting ♂β into ♃β'.

If, however, one does choose to fix ♂β inside ♃β' (or, what is the same, to amalgamate ♂α and ♃β') – the only way of arriving at a 49 sphere cosmos – this already constitutes a *fundamental* modification of the

110 p. 217.
111 p. 391.

original Eudoxian theory. As one seeks to make Aristotle's cosmological system work, even for him, it appears that his synthesis must become not simply a compilation of existing techniques, but a substantial contribution to ancient astronomical theory (which, in other ways, it surely is). The 49 sphere account forces the historian towards that conclusion. But there is another solution to this perplexity; it is more 'Aristotelian' and more in keeping with the orthodox view of how The Philosopher adopted and adapted Eudoxos and Kalippus.

The idea that $\delta\alpha$ should be a sphere separate from $\tfrak{2}\beta'$, yet have no local motion of its own, is wholly 'un-Aristotelian'. The fusion of $\delta\alpha$ and $\tfrak{2}\beta'$ is to be preferred, according to Aristotle's principles, to any rigid bracketing of the two different spheres.[112] The innermost $\delta\alpha$, lacking a local motion because of the bracketing, will not be governed by an unmoved mover, or principle, and would thus not be a perfect substance.[113] By this alone it appears that $\alpha$ must be omitted, for each planetary sphere-cluster, if Aristotle's cosmology is to succeed. Maintaining each $\alpha$, but denying it any local motion, clashes with all other Aristotelian cosmological dicta.

Now it might be urged that 55 spheres is all right if only we assume that each new $\alpha$, in each subsidiary planetary sphere-cluster, has *absolutely* the motion of the fixed stars. One could suppose that $\alpha$'s motion began anew for each planetary set. But this is incompatible with Aristotle's desire to set out a single unified, mechanically-articulated cosmology. He describes the motions of the spheres, local and absolute, in a wholly uniform way. Just as the absolute motion of B in Figure 23 is affected by that of A above it, and affects the motion of C below it – *so the motions of $\delta\alpha$ should be affected mechanically by the motions of the sphere above it* ( $\tfrak{2}\beta'$), *just as it should affect those of* $\delta\beta$ *below*. When it is suggested that 55 spheres will work if $\tfrak{2}\beta'$ and $\delta\alpha$ (although independent and separate 'unmoved movers') both have the *absolute* motion of the fixed stars – we are being told that there is *no mechanical connection* between $\tfrak{2}\beta'$ and $\delta\alpha$. This is contrary to Dreyer's, Ross' and Heath's interpretation of Aristotle and contrary also to Aristotle's stated objectives. If there *were* such a mechanical connection, however, the result would be as with

---

[112] For, as Ross says: "Every perfect substance *produces a motion* in the heavens", p. 394, my italics.
[113] Yet the number of these unmoved movers Aristotle clearly fixes at 55.

all the other rolling and unrolling spheres; the moon would whirl around us NASA-style. If, on the other hand, the moon is slowed down to its actually observed angular velocities, and the 55 spheres are *still* insisted upon – all mechanical connections between the sphere-clusters being broken thereby – then Aristotle's cosmology becomes a transparent reiteration of what Eudoxos and Kalippus had set out earlier. These earlier thinkers imagined no mechanical, or even kinematical, connection between any two of their planetary sphere-clusters. Aristotle apparently differs in this respect: his was meant to be a unified, mechanical adaptation and synthesis of the Eudoxos-Kalippus work. His work was thus theoretically different from theirs. But again, if Aristotle's cosmic mechanization is taken literally, and the 55 spheres are yet preserved, then the moon will end up with rapid motions we never observe, as was just demonstrated. As we saw, if the moon is held back to fit the observations, then any mechanical linkage between sphere-clusters simply snaps; Aristotle's entire cosmology becomes a childish re-scaling of the Eudoxian calculation technique, with philosophical decoration in abundance.[114]

The last possibility is that the moon should, of course, be held to its observed motions; the spheres ♄α, ♃α, ♂α, ☿α, ♀α, and ☉α should have a local motion on their own axles (in accordance with other Aristotelian principles); the Eudoxos-Kalippus technique should remain theoretically unmodified; and Aristotle's essential contribution, viz. a quasi-*mechanical* connection between all the nested planetary sphere-clusters, should be realized to the full. Granted these as the controlling conditions of our interpretative study, *no alternative now exists but to change the*

---

[114] Strangely enough, Heath actually embraces this interpretation:

"Of the several spheres which act on any one planet, the first or outermost alone (i.e., ♄α, ♃α, ♂α ... etc.) is moved by its own motion exclusively ..." (*op. cit.*, p. 230, my insert).

In support of this Heath cites *Physica*, VIII, 6, 259b 29–31, which is not really relevant so far as I can see. For reasons given earlier, this remark of Heath's is wholly unacceptable. It is even at variance with his own position in other places: "... the world ... derives its eternal motion from the *primum movens* ..." (p. 228); "the *primum movens* operates on the universe from the circumference ..." (p. 226). Compare *Physica*, VIII, 10, 276b, 7–9. Either the *primum movens* drives the entire planetary machinery 'from the circumference', or else each planet's α sphere 'is moved by its own motion'. If the latter, then Aristotle's cosmology is not integrated and mechanical. If the former, then 55 spheres will generate counterfactual expectations.

*number of spheres* (and associated 'unmoved movers'), *from 55 to 61*. This change consists not in omitting 6 spheres, as an army of commentators has urged, but in adding 6. In this way Aristotle's objectives may be most readily achieved, while yet doing no violence to the stellar contributions of Eudoxos and Kalippus.

Ross paraphrases Aristotle as follows:

We must suppose for each of these bodies except the moon, counteracting spheres, one less in number than the positive spheres, to neutralize their action on the outer sphere of the next system (counting inwards).[115]

Sir David continues:

Eudoxos and Kalippus had offered a purely geometric account of the planetary system; Aristotle aims at a mechanical account, and cannot isolate the system of one planet from that of the next. He therefore supposes for each 'planet' except the moon certain spheres which 'roll back' the outer sphere of the planet just nearer to the Earth than the given planet, i.e., which prevent the influence of the forward-moving or deferent spheres of one planet from affecting the next.[116]

In short, Aristotle turns Eudoxos' *axes* into *axles*. Geometrical axes, being one-dimensional, can be 'fixed into' geometrical points: physical axles, being three-dimensional, cannot. As moves the bushing, so moves the axle fixed therein (whether the latter be in independent motion or not). Aristotle's cosmology is a timepiece of turning axles set into moving bushings. The result is an increasingly composite motion, unless this be progressively and completely nullified by unrolling axles – which $\beta'$ *never* succeeds in bringing about.

The spheres $\beta'$ (the innermost of any system) fail to neutralize the effect, e.g., of Mars' sphere system upon that of Mercury (to adopt Aristotle's ordering). Complete neutralization could be achieved, while preserving everything else Aristotelian, only by *adding* to each sphere system one more unrolling sphere, $\alpha'$: the effect of this would be to cancel the diurnal rotation of the outermost sphere ($\alpha$) of each system. Each new subordinate system, then, would begin from absolute rest. Any point within $\alpha'$ will be at rest in absolute space. Hence those two points in which $\alpha$'s axle will be fixed also *rest* in absolute space.

So now the new $\alpha$ will have just those motions required of it. Its local motion = one axial rotation every 24 hours. (Its motion in absolute

---

[115] Aristotle's *Metaphysics*, vol. 2, p. 383.
[116] *Op. cit.*, p. 391.

space = exactly that of the diurnal rotation of the fixed stars.) Thus if each
α is to do what Aristotle wants of it, there must be 'above' it an α' which
will place α's axle in *absolutely* fixed and motionless bushings.

This is clearly what Aristotle wants; it is provided here within the
unified and integrated kinematical machinery he is ambitious to in-
stantiate. By adding one such new sphere (α') to each planetary system
– to Saturn, Jupiter, Mars, Mercury, Venus, and the sun – we will be
adding 6 spheres. The total: 61. This suggested modification has advantages not apparent in any other alternative. The 'unrolling' technique
now operates just as Aristotle describes it; the basic Eudoxian scheme is
preserved since the first sphere (α) of each system now has both a local
motion *and* an absolute motion corresponding to that of the fixed stars.[117]

α now rotates on its own axis-axle, as required both by the Eudoxian
geometrical technique and by Aristotle's 'physical' doctrine of the unmoved movers. For these reasons the 61-sphere cosmos appears to be
the one Aristotle *must* have intended – in principle, at least. Had he but
worked his composite cosmology clear through, he would have seen
(more quickly than we have) that 55 spheres cannot work, and that 49
spheres would constitute a modification of Eudoxos' astronomy so
serious that Aristotle could not consciously have attempted it without
some signal to that effect.

Two connected points arise at once. One of these has been much discussed, the other not at all. They affect the question of how many spheres
Aristotle had to have in his cosmology in order to generate observationally acceptable results. We will consider them briefly.

It was noted in the 19th century[118] that Aristotle's account of shooting
stars and comets necessitates unrolling spheres *beneath* the moon.[119]
Such phenomena were said to result from terrestrial 'exhalations' rising
to the sublunary sphere, where they contact a 'warm, dry substance'.
The exhalations are there kindled and carried around atop the sublunary
sphere. But large comets move with the fixed stars as if they shared their
diurnal rotation. Hence some sphere below the moon must have the

---

[117] These it cannot have in an 'economical' 49 sphere cosmos, wherein α (having been
fused with the β' above it) has an *absolute* motion corresponding to that of the fixed stars,
but a wholly different local motion, namely – that of β'.
[118] Martin, *Mémoires de l'Acad. des inscr. set. Belles-Lettres*, XXX, pp. 263–4.
[119] Cf. *Meteorologica*, I, 6–8, pp. 342b. ff.

motion of the fixed stars. At this juncture Heath makes an obscure statement:

The four inner spheres producing the moon's own motion should therefore be neutralized as usual by the same number of reacting spheres.[120]

For one thing, Aristotle requires *five* spheres initially to produce the moon's motion, not four. This much comes straight out of Kalippus. But here is the one case where we do not want the unrolling spheres completely to neutralize the effects of the rolling spheres. We want to give 'terrestrial exhalations' like comets and the Milky Way the fixed star's motion. This means we would *not* add to the rolling spheres the same number of reacting spheres (5 in this case); if we did that everything beneath the moon's spheres would be at absolute rest. That might be all right for evanescent phenomena like meteors and the Northern Lights, but comets and the Milky Way persevere, and must therefore share in the diurnal rotation of the cosmos. Hence, no sphere answering to $\mathbb{D}\,\alpha'$ should be introduced beneath the moon. Any point on $\mathbb{D}\,\beta'$ might have been fixed into the lunar system's outermost sphere $\mathbb{D}\,\alpha$ which itself has the diurnal rotation – locally and absolutely.

So (following Martin's intuition), it appears that the *Meteorologica* would require four more sublunary 'unrolling' spheres – $\mathbb{D}\,\zeta'$, $\mathbb{D}\,\delta'$, $\mathbb{D}\,\gamma'$, and $\mathbb{D}\,\beta'$. This will again raise our total now to 65.

A more serious issue, as yet wholly undiscussed in the literature, concerns Saturn's outermost sphere $\hbar\alpha$, as against that of the fixed stars. Earlier considerations arise anew. Is $\hbar\alpha$ identical with the sphere of the fixed stars?[121] Or, does $\hbar\alpha$ just have the *same motion* as do the fixed stars? Certainly Aristotle nowhere identifies Saturn's $\hbar\alpha$ sphere with that of the fixed stars. Nothing in Eudoxos' theory distinguishes $\hbar\alpha$ in this way either – by identifying it with the stars. Nothing prevents this identification. To have done this however, would have set Saturn apart as the only planet literally possessing the fixed stars as its 'own unmoved mover'. Surely this stellar Saturnian distinction would have merited some small comment by The Philosopher?

Suppose, however, that Saturn is not distinguished thus. Suppose that $\hbar\alpha$ is not identical with, but only fixed within, the sphere of the fixed

---

[120] *Op. cit.*, p. 219.
[121] Ross denies this; *op. cit.*, p. 387, 18.

stars. Our original perplexity breaks out all over again. If there is any mechanical connection between Saturn's sphere α and that of the fixed stars, the result will be that any point on ♄α will move with *twice* the angular velocity of any fixed star; ♄α will be spinning around its own physical axle, and the axle itself will be fixed in the stellar sphere, and hence will also have a diurnal rotation of its own, independently of what the sphere ♄α is doing *around* it. This, plus other accretions already discussed, would have the ultimate effect of whirling the moon 'round the earth eight times a day, or once every three hours.

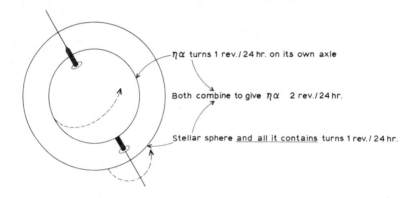

What is the alternative? If the two spheres (the stellar sphere and ♄α) are kinematically linked at all, as Aristotle wishes them to be, then there must be an unrolling sphere immediately below the fixed stars and immediately above Saturn's ♄α. This new substellar sphere would spin once every 24 hours from west to east; any point on it would be at rest in absolute space. Aristotle may then fix Saturn's outermost sphere, ♄ α, into this substellar sphere and treat it in the standard Eudoxian manner. This new substellar sphere – let us designate it 'Σ'' – will raise the total number of spheres to 66.[122]

But no matter how one tallies the number, an aggregate of 55 spheres *simpliciter* cannot be correct. This figure is objectionable not because it comprises redundant spheres, or because it is geometrically inelegant –

---

[122]  61, as urged earlier, plus 4 lunar 'unrolling' spheres, plus this substellar sphere. Or if, despite our cautions, one still opted for the 49 sphere arrangement in the tradition of the Commentaries, the new tally would be 54 spheres; 49 plus 4 lunar 'unrolling' spheres, plus 1 substellar sphere Σ'.

these being the usual reasons advanced (for two millennia). It is objectionable because, for Aristotle's purposes, it is just false.

Why such detail about what may seem but a minor feature of a crude
version of a disconfirmed cosmological theory? The answer is simple:
the reader who has come this far with me in counting Aristotle's spheres
now really understands his planetary conception – its objectives, its
assumptions, its formal machinery and its points of vulnerability. These
are the advantages in working out The Philosopher's cosmology in such
close order. Could there be a more satisfactory *denouement* for Part I
of Book One?

Philosophers and historians of science have dwelt on the artificial
cumbersomeness of this rococo Aristotelian cosmological system.
Thus: "... it is no wonder that philosophers after him found his
machinery rather cumbersome".[123] But such scholars may have been
hasty. For here is a remarkable cosmological system. Aristotle supplies
in an elegant and economical way, what was lacking in Eudoxos and
Kalippus.

Of course, he could not cope with questions concerning the actual
distances of celestial bodies from the earth, or from each other. But
neither could his predecessors, contemporaries, or immediate successors;
nor could any other astronomer before the 16th century. Aristotle – like
Eudoxos, Kalippus, and (later) Hipparchus, Apollonios and Ptolemy –
was phenomenological in his studies of the cosmos. His problem was to
discover a possible *rationale* behind the circlings and wanderings of those
points of light in the heavens, a rationale more satisfying than that engendered by Eudoxos. These lights might have been generated at almost
any distance from the Earth, and from each other. The only restriction
was that they could not be infinitely far from us. For, as we saw, a radius
of infinite length could not possibly swing 'round us in any finite time,
much less in a mere 24 hours.[124]

If the fixed stars are but a finite distance from us, this must necessarily

---

[123] Dreyer, *A History of Astronomy*, p. 113.
[124] *De Caelo*, II, Chap. 4. Again one cannot but be fascinated by this argument. If we are
at a finite distance from the stars, then any motion on our parts should lead to some
detectable parallactic aberration in our successive plottings of those stars' positions.
The only way for the earth to have a motion and yet stellar parallax remain unobserved,

be so with the planets, which are often observed to eclipse the stars; Aristotle himself saw the moon cover Mars.[125]

So the universe is a nest of perfect, finite spheres; it is enclosed within the most perfect of them all, whose "Circular motion is necessarily primary ... the circle is a perfect thing ... the heavens (thus) complete their circular orbit ... [and] the heavens ... must necessarily be spherical".[126] "Circular motion ... is the only motion which is continuous".[127] "It is in circular movement ... that the 'absolutely' necessary is to be found."[128] And since the outer sphere moves perfectly, it must therefore move the more quickly; the quickest is most perfect.[129] Here then is the basis of all order in the universe. For the motive power behind this stellar sphere is the primary cause of motion, the 'primum mobile'.

The sphere of the fixed stars *is that which moves all other spheres*, and that which is placed beneath this and has its movement in the circle which bisects the zodiac is common to all.[130]

Its power extends, as was intimated earlier, from the circumference to the center.

The stars are at rest in their spheres, and only the latter are in motion.[131]

The sphere's motions are completely uniform. If they were not, it would

---

is for the stars to be infinitely distant – a possibility already ruled out by the first condition. Therefore the earth *cannot* move.

As a cogent argument from what is actually observed, this must rank high in the history of thought. Aristotle dovetails the two undeniable celestial observations of his time – (1) the diurnal stellar rotation, and (2) the absence of any detectable parallax – with two powerful principles: (a) the impossibility of an infinitely-distant object circling us in a finite time, and (b) the necessity of an observer detecting parallax amongst the fixed objects past which he travels. A diurnally spinning earth is blocked by the second principle (b): no stellar parallax. Could this be due to an infinite distance of the stars? No; that possibility is blocked by the first principle (a): stellar rotation in a finite time. There is a slight *petitio principii* here, but no matter. After Aristotle the idea that the stars are fixed into a sphere of immense extent although finite (and hence beyond practical detection) was almost universally accepted. Among the Greeks Herakleides and Seleukus were alone in rejecting it, and Cicero and Plutarch stood against it amongst the Romans. (Cf. Seneca, *Divinatione* II, 43, 91. 'Infinite and immense' are the operative terms here in Seneca; cf. also Plutarch, *On the Face in the Disc of the Moon*, where it is argued that the earth is not in the center of the universe, since space is infinite and hence has no center!)

[125] *De Caelo*, II, 12, p. 292a.
[126] *De Caelo*, 268b, 276b.
[127] *De Generatione et Corruptione*, 337a.
[128] *Op. cit.*, 338a. Compare also *Physica* (223b, 227b, 265b, 248a, 261b, 262a).
[129] Another singularly revealing inference.
[130] *Metaphysics*; my italics.
[131] *De Caelo*, II, 13, 293b.

be because the primary power waxes and wanes; that is absurd. Furthermore, while terrestrial elements are endowed with rectilinear motion – as fire moves 'upward' towards the circumference, and earth moves 'downwards' towards 'the center' – celestial bodies, being primitive and superior, must have the superior motion, viz., in a perfect circle.[132] Indeed, they are *observed* to have this motion, an empirical confirmation of the entire argument preceding. Finally, all stars encircle the earth with just the one velocity of the stellar sphere itself. Thus the stars are at rest, imbedded in that sphere.

… there are two motions of a spherical body, rolling along and whirling [i.e., traveling and twisting, translation and rotation], then the [individual] stars, if they have a motion of their own, ought to move in one of these ways. But it appears that they move in neither of these ways. For if they whirled (rotated), they would remain at the same spot and not alter their position [again, singular!], and yet they manifestly do so, and everybody says they do.[133]

This seems to allow a sphere to twist, *or to travel, but not both!*[134] Thus Aristotle turned the acute mathematics of Eudoxos into a cosmic, spherical physics. The spheres in the geometer's abstract imagination were drawn out, enlarged, and crystallized. They were nested within one another and spun on myriad axles. As a natural philosopher Aristotle wished to explain appearances by an account of nature as it really is. He was not, like Eudoxos, content to enumerate possible structures nature might have. Aristotle sought a systematic cosmology, not an accurate astronomy; his beautiful picture of nested spheres (Figure 24) could hardly be expected to enable prediction of the complete dispositions of the heavenly bodies at given times.

This profound difference in attitude between the mathematician and the ancient physicist is explained in *Physica* where Aristotle writes:

The next point to consider is how the mathematician differs from the physicst … is astronomy different from physics or a department of it? It seems absurd that the physicist should be supposed to know the nature of sun or moon, but not to know any of their essential attributes, particularly as the writers on physics obviously do discuss their shape also and whether the earth and the world are spherical or not. Now the mathematician, though he too treats of these things, nevertheless does not treat of them as the limits of a physical body; nor does he regard the attributes indicated as the attributes of such bodies. That is why he separates them; for in thought they are separable from motion, and it makes no

---

[132] This argument is mirrored in Galileo, *Dialogue Concerning the Two Chief World Systems*; cf. also *De Caelo*, I, 2, p. 269a and II, 7, p. 289a.

[133] *De Caelo*, II, 8, p. 290a.

[134] Compare Euclid, p. 7 *supra*. Thus the moon was said not to twist, but only to travel 'round us – for is not but one of its sides always facing us?

difference, nor does any falsity result if they are separated ... the more physical of the branches of mathematics, such as optics, harmonics, and astronomy, ... are in a way the converse of geometry. While geometry investigates physical lines but not *qua* physical, optics investigates mathematical lines, but *qua* physical, not *qua* mathematical .... How far then must the physicist know the form or essence? ... the physicist is concerned only with things whose forms are separable indeed, but do not exist apart from matter ...[135]

The point of this passage is sharpened by Geminus, in the First Century B.C. There had grown by that time, a considerable tradition of astronomers of Eudoxos' outlook, rather than Aristotle's. But Geminus argues that physics examines the nature, power, quality, birth and decay of the heavens – the better to understand and explain what *kind* of thing the universe is. Astronomy seeks only to discover the arrangement of the celestial bodies, their figures, sizes and distances, the eclipses, conjunctions and movements of the stars, the better to plot and predict the places at which the planets and other points of celestial brilliance are to be found. Astronomy is sustained by arithmetic and geometry, the formal 'computers'. But the physicist (i.e., the natural philosopher, the student of $\phi\upsilon\sigma\iota\sigma$) seeks causes and moving 'forces' – about which mathematics *per se* is completely silent. The astronomer seeks methods of calculation which, when accepted and used, could locate and describe the observed phenomena, just as a geometer 'describes' a circle. Geminus asks:

... why do sun, moon and planets appear to move irregularly? Because when we assume their orbits to be excentric circles, or the stars to move on epicycles, the appearing anomaly can be accommodated; and it is necessary to investigate in how many possible ways the phenomena can be represented, so that planetary theory may be made to agree with the aetiology in an admissible manner ... In general, it is not the astronomer's business to see what by its nature is immovable and of what kind the moved things are, but framing hypotheses as to some things being in motion and others being fixed, he considers which hypotheses are in conformity with the phenomena in the heavens. He must accept as his principles from the physicist, that the motions of the stars are simple, uniform, and regular, of which he shows that the revolutions are circular, some along parallels, some along oblique circles.

And compare:

... astronomy ... but proves the arrangement of the heavenly bodies ... it tells us of the shapes and sizes and distances of the earth, sun, and moon, and of eclipses and conjunctions of the stars, as well as the quality and extent of their movements ... The Things ... of which alone Astronomy claims to give an account it is able to establish by arithmetic and geometry ... the astronomer ... is not qualified to judge of the cause ... he invents by way of

---

[135] *Physica II*, 2, 193b, 194a, 194b.

hypothesis, and states certain expedients by the assumptions of which the phenomena will be saved ... [136]

Aristotle and Geminus both distinguish the physical explanations of celestial phenomena from mathematical hypotheses which merely predict forthcoming phenomena in a formal manner. The astronomer's task is to frame a calculus which can represent, e.g., the retrograde loops of Mars and Venus – thereby making them subject to calculation and forecasting. It does not matter whether the associated theory is physically true, or false, or even without a complete physical interpretation.

This distinction closely parallels the one basic to this book – that between explanation and prediction. Successful prediction just by itself (i.e. perhaps 'by accident') could not have been enough for Aristotle. Geminus seems more flexible. And one can peer further back with 20/20 hindsight to detect something of this same attitude in Eudoxos and Kalippus, both of whom would have maintained the distinction, while rejecting as beyond utterly human powers any attempt at an explanation of celestial phenomena. This is explicitly stated by Ptolemy; both approaches develop side by side until the time of Newton. Thus, to anticipate:

Thirteenth-century astronomy was ... concerned mainly with a debate as to the relative merits of physical as compared with mathematical theories in accounting for the phenomena. The former were represented by Aristotle's explanations, the latter by Ptolemy's ... [137]

And we shall see in Book Three that Newton-the-natural-philosopher was at last successful in combining calculational astronomy with discursive physical theory – forecasting power with understanding; prediction with explanation.

Another related contrast thus flows here beneath the linguistic surface of the works of Aristotle and Geminus. It is that between computational *astronomy* and philosophical *cosmology*. Virtually all thinking about the universe before Eudoxos was pure cosmology, or, as an astronomer today might say 'mere philosophy'. It consisted in 'explaining' celestial phenomena by imagining them within complete and unified pictures of the cosmos. The 'worlds' of Plato and his predecessors provided intelligible patterns whose constituent pictorial descriptions of pheno-

---

[136] See Heath, *Greek Astronomy*, pp. 124–125. (*Editor's note.*)
[137] Crombie, *Augustine to Galileo*, p. 56.

mena made them seem less surprising than before. Were these word pictures 'explanations'? Was Figure 12? Certainly. To deny this would be to legislate on how 'explanation' *must* be used; some philosophers in the Popper-Hempel tradition do just this. But they leave undiscussed thereby those accounts which most often do in fact *count* as the offering of explanations. The cosmologies tendered by the Ionians, the Pythagoreans, by Plato and Aristotle these may have been inadequate. But they were nonetheless explanations, albeit false ones. When a prediction fails, we do not say *it* was never a prediction at all. Why such different treatment for explanations, which are supposed to be so similar in logical structure: why not 'false explanations'?[138] When an explanation falls short of Hempel's full analytic demands, was it never an explanation at all? An explanation may fail: it might be inadequate, unilluminating and ill-founded. But this does not mean that it was never an explanation for anyone – that it never did what explanations are meant to do when offered. That judgment would force a rewriting of all intellectual history; for, since their accounts would not suit us today, we should have to deny that Hipparchus, Ptolemy, Copernicus, Kepler, Newton and Leverrier ever offered explanations at all: *reductio ad absurdum*. Ancient cosmological explanations indeed fell short of the ideal situation described in our *Introduction*. Few predictions followed from these speculative pictures of the universe. These were, nonetheless, genuine cosmological explanations which were inadequate for prediction.

Eudoxos was somewhat different in all this. Explanation was still his ambition – in Dray's sense of explaining α by showing a possible (and intelligible) formal structure for α. Eudoxos was too good a mathematician to be overruled by a desire for wholesale intellectual quiescence. He knew that difficult problems could not be solved wholesale, by single, comprehensive, cosmological intuitions. They had to be fragmented and decomposed into simpler, constituent problems. Eudoxos appreciated that reconciling perfectly circular motion with the planetary loopings would be supremely difficult. So, rather than mappping the universe onto a single sheet, he engaged the planets one at a time. He sought individual accounts for individual anomalies. He sought some intelligible structures

---

[138] When an enthusiast predicts the end of the world tomorrow and is wrong, will we deny that he made a prediction?

behind the errant wanderings of the planets; but not by impressing some super-scheme upon them 'from above'. He knew that until he could describe the planetary meanderings in detail, he would not have described them at all. This spirit breathes in Kalippus and dominates Ptolemy – who is, by Geminus' criterion, the compleat astronomer.

Aristotle pressed this first leaf of mathematical astronomy between the pages of his weighty cosmology. His ingenious systhesis was the first comprehensive, and yet detailed picture of the universe. Aristotle explained both the general plot and the particular acts and schemes of the cosmos. But he added nothing to predictive astronomy.

Still, Aristotle's cosmology is a better *scientific* undertaking than philosophical or historical scholars seem yet to have wished to concede. Granted, there is much metaphysical hocus-pocus in *De Caelo* and the *Metaphysics*. Nonetheless, considered within the context of 400 B.C., it resembles the endeavors of the systematizers of the 17th century who were, after all, not above a little hocus-pocus themselves. (We are not above our fair share today.) Aristotle discovered no new cosmological or astronomical facts. He never claimed to have done so: how like Newton! But he was fully conversant with *all* the data known to the astronomers of his day: again, how like Newton – and how unlike Aristotle's modern detractors. It was Aristotle's objective to find some unified framework within which all the then-extant observations, and the best available theories, could be integrated and harmonized intelligibly. To deny the term 'scientific' to such an undertaking would be to restrict science to fact-finding, by which some think first of 'bug hunting'. But then the synthesis of Copernicus would suffer in appraisal also. He, like Newton after him and Aristotle before, revealed no new data, nor did he seek any. Such thinkers sought to recast old facts and theories into new scientific frameworks – new patterns, new structures. Much of what Aristotle said is false. His scientific endeavors may have been overruled by myopic metaphysics. But that he, like Copernicus and Newton, sought an intelligible framework for the many well-established facts known in his day seems indisputable. If Aristotle's theory is somewhat deficient in predictive power (it surely was!) it could not possibly have been more so than were the calculating devices of Eudoxos and Kalippus, both of whom usually get good marks in history of science. Aristotle sought only to accommodate their 'geometric computers' within his larger cosmological picture. It is not too far-fetched to see in this something like Einstein's harnessing of Reimann's techniques for the larger objectives of general relativity.

Physics and Mathematics; explanation and prediction; cosmology and astronomy: these dichotomies were sharpened by the force with which the idea of regularly circling stars clashed with the concept of freely wandering planets. Our two great facts of the heavens, by their abrasive contact within the Greek mind, make the subsequent history of astronomy seem like little more than a perpetual resounding of 'Plato's question'. Claudius Ptolemy grinds these facts somewhat and brings them a little closer together. He was a fine mathematician, and a careful student of Aristotle's *Physics*. Not only does he predict celestial move-ments with unprecedented accuracy, but in Peripatetic prose he parleys for just a modicum of intelligibility and understanding of the heavens. Ptolemy was an ingenious and skillful astronomer, a man who could predict with his calculating devices, what *will* happen in the heavens tomorrow. He was also a geocentric cosmologist in the best Aristotelian tradition, a man who could explain, by recourse to The Philosopher's Metaphysics, what *had* happened in the heavens yesterday. In Ptolemy many turbulent streams of ideas and cross-currents of thought are made confluent and laminar. If these ideas never really intermingle and coalesce, but only run parallel to each other, it is not for any lack of wit and industry in that distinguished Alexandrian.

# BOOK ONE

## PART II

Ptolemy and Prediction

Four worthies are usually cited, within astronomical histories, as subsequent to Aristotle and prior to Ptolemy. These are Herakleides (4th century B.C.) and Aristarchus (fl. 281, B.C.), Apollonios and Hipparchus (3rd, and 2nd century B.C.). How do they further our study of the Explanation-Prediction contrast as it stalks through the History of Planetary Theory?

Herakleides explained the apparent diurnal rotation of the heavens by supposing the earth to turn daily on its own axis, the heavens remaining at rest.[1]

Simplicius says:

Herakleides of Pontus, assuming the earth to be in the middle and to move in a circle, but the heavens to be at rest, considered the phenomena to be accounted for.[2]

Chalcidius remarks further that:

Herakleides of Pontus, in describing the path of Venus and of the sun, and assigning one mid-point for both, showed how Venus is sometimes above and sometimes below the sun.[3]

Compare Vitruvius:

The stars of Mercury and Venus make their retrogradations and retardations around the rays of the sun, making a crown, as it were, by their courses about the sun as center.[4]

And Martianus Capella:

Venus and Mercury ... do not go around the earth at all, but around the sun in freer motion.[5]

Aristarchus asserted still further that the earth moves in an oblique circle around a fixed sun.[6]

---

[1] Cf. above, pp. 16; compare Aetius III, 13.3 (page 378 in Diels).
[2] *De Caelo* II (Heiberg), p. 519.
[3] *Commentary on Plato's Timaeus*, p. 109.
[4] *On Architecture*, IX, 1.6.
[5] From *On the Marriage of Philology and Mercury*, VIII.
[6] Thus Archimedes says:
   "Aristarchus of Samos' ... hypotheses are that the fixed stars and the sun remain unmoved, that the earth revolves about the sun in the circumference

Plutarch remarks:

Cleanthes held that Aristarchus of Samos ought to be accused of impiety for moving the heart of the world.[7]

From Plutarch also [8] it seems that Seleukus adopted Aristarchus' heliocentric hypothesis, not as a mere hypothesis for calculation, but as a brute fact:

[Does the earth rotate?] as Aristarchus and Seleukus have afterward shown, the one supposing it only, but Seleukus affirming it as true.[9]

Sextus Empiricus wrote:

Those who do not admit the motion of the world [i.e., the heavens] and believe that the earth moves, such as the followers of Aristarchus the mathematician, are not hindered from discerning the time.[10]

But, as was argued earlier (pp. 18–21), the influence of these two natural philosophers on the development of astronomical thought was not great – justifiably not, perhaps. Aristarchus' doctrine conflicted with the metaphysical principle of the four elements' *natural places*, earth *necessarily* being central – i.e. 'furthest down'. It ran counter also to the then-standard alternative, the Pythagorean idea of a central fire which, by Aristarchus' time had assumed the authority of quasi-religious dogma.[11] The hypotheses of Aristarchus and of Herakleides were rejected therefore. For if the earth really moved – really rotated or revolved – the fixed stars should appear to group and regroup, like the columns of the Parthenon as we walk before them. The stars do no such thing. Nor could the seasonal inequalities be squared with Aristarchus' central placement of the sun. Empirically, then, these explanatory hypotheses failed. Had such brilliant conjectures any real influence at

---

of a circle which lies in the midst of the course of the planets, and that the sphere of the fixed stars, situated about the same center as the sun, is so great that the circle in which he supposes the earth to revolve bears such a proportion to the distance of the fixed stars as the center of the sphere bears to its surface."

*The Sand-Reckoner*, I (in Heath, *The Works of Archimedes*, p. 222).

[7]  *Op. cit.*, Section 6.
[8]  *Platonic Questions*, 1006c.
[9]  Compare Aetius III. 17.9.
[10]  *Adversus Mathematicos*, X 174.
[11]  Cf. Plutarch, *De Facie in Orbe Lunae*, 922F–923a.

all, it was only many centuries later – as when Copernicus notes them to excuse his own heterodoxy. So Herakleides and Aristarchus will not be disinterred further in this work: neither was a 'Copernicus of Antiquity'. The mathematico-astronomical-predictional legacy of Aristotle fell directly to two heroes within history of science – Apollonios (c. 230, B.C.) and Hipparchus (c. 130, B.C.). But the legacies were profoundly altered in the conveyance.

The very primitive theory that heaven consisted of a *single* sphere, this was smashed by observing the sun, moon, and planets slowly inching from west to east *against* the east-west diurnal motion of the stars. The less primitive theory, that these sub-stellar bodies were themselves fixed into smaller spheres, counter-rotating (or just moving more slowly) but co-axial with the celestial sphere – this theory was dented and bruised by the astounding fact that the planets looped backwards in their orbits. The ingenious computational scheme of Eudoxos – as generalized and crystallized into a cosmology by Aristotle – was in its turn cracked and battered by (1) the recalcitrant inconstancy with which the planets reversed their orbits, and (2) by enormous variations in planetary brightness, especially Venus, suggesting thereby a variable distance from the earth. And consider 'Halley's Comet', as it actually moved through our solar system in the Fourth and Third Centuries, B.C.

This spectular object was not known by the ancients to be circumsolar. Rather they supposed it to be a kind of 'terrestrial exhalation'. Nor could they imagine that it was one and the same object which appeared both in 340 B.C. and again in 265 B.C. However, it was by Tycho Brahe's time too obvious from its trajectory that any such comet would have to intercept, and shatter through, the apparently solid spheres on which Aristotle supposed the planets to be carried.[12] No such crystal shattering cataclysm was ever observed; this made the hypothesis of actual spheres, as contrasted with Eudoxian imagination-props, highly untenable to Tycho and all subsequent astronomers. But to many pre-Ptolemaic observers the comets were not celestial bodies at all; they roam at random all over the heavens.

Seneca muses that there is no conclusive reason for thinking the few

---

[12] Cf. *Phenomanae Recentioribus* ... Book VIII. It is not certain that Aristotle really took the celestial spheres to be substantial crystal bubbles: but some of his followers certainly did.

planets we do know of to be the only ones; there may be others which are generally invisible, because their circles are placed so that they can be seen only when they pass to us [13]. Clearly then, comets did not slip easily

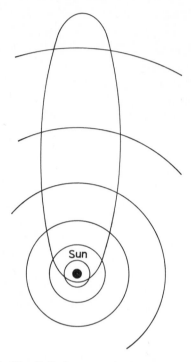

Fig. 25.   Halley's comet – Ancient appearances.

(*Editor's note:* this caption was struck out by the author in the copies of the manuscripts found among his papers, but is included here because no alternative caption was substituted for it in the manuscript.)

into any Eudoxian sphere inside Aristotle's cosmological machine.

Again, though the relatively regular reversals of Mars might just be 'accounted for' by judicious jockeying of Kalippus' nested spheres, it is difficult to visualize how *any* finite combination of spheres (much less

[13] *Quaestiones Naturales* VII, 13. How darkly prophetic of Uranus, of Ceres, of Neptune, of Pluto, and of ....

one fixed configuration of the same ones) could explain such things as Venus' successive paths.[14]

The deathblow to the Eudoxian-Aristotelian astronomy-cosmology, however, came from noting how enormously the brightness of Mars and Venus vary from one time to another. We now correlate this variation with changes in the apparent diameter of the planet. Venus, for example, varies in size somewhat as in Figure 26.

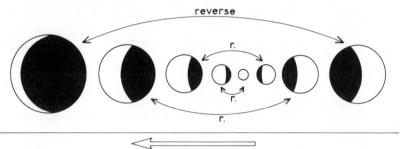

Fig. 26.   Variations in size and brightness of Venus.

These phases cannot be detected without a telescope. Naked-eye astronomers of Greece encountered all this simply as a dramatic change in brightness, which was inconsistent with any geocentric, spherical model of the cosmos. A planet at constant radial distance from us could not be so variable in its appearance.[15]

If the cosmic motions were to be understood, then, the homocentric theory of Eudoxos, and its Aristotelean generalization, could not be the way to proceed. The Philosopher had misgivings all along:

Aristotle ... discusses objections to the hypotheses of astronomers arising from the fact that even the sizes of the planets do not appear to be the same always. In this respect Aristotle was not altogether satisfied with the revolving spheres, although the supposition that, being concentric with the universe, they move about its center, attracted him. Again, it is clear from what he says in Book *Λ* of the *Metaphysics* that he thought that the facts about the movements of the planets had not been sufficiently explained by the astronomers who came before him or were contemporary with him.[16]

But was explanation (Aristotle's kind) the only worthwhile goal for the astronomer?

[14] Cf. Figures 10 and 11 *supra*. Of course, as a matter of formal principle *some* finite constellation of spheres could generate *any* resultant line (cf. pp. 99–117 *infra*). But the algebraically-unaided imagination is all but paralyzed in any such practical confrontation.
[15] Cf. Sosigenes, in Simplicius, *De Caelo*, pp. 504–505 (Heiberg).
[16] Sosigenes, *ibid*.

After Aristotle, practical-observational astronomy underwent intense development: men like Herakleides, Aristarchus, Archimedes, Apollonios and Hipparchus poked their heads through the philosophical clouds and looked straight up to the stars – natural objects to be comprehended in themselves, not just 'asides' in some noble metaphysical undertaking. Accurate observing had been going on all along, especially in the Near-East, in Egypt, and in Greece [17]. The industrious and clear-sighted had for ages been plotting the planetary positions. They had drawn up tables summarizing and tabulating these observations; they had wrestled with calendrical problems, and had ventured an occasional prediction – of an opposition or conjunction perhaps, or an occultation – or even of a solar eclipse. For men of such a turn of mind, undisciplined cosmological speculation held few attractions. [18]

Fruitless attempts had multiplied attempts to explain the celestial machinery wholesale; while the everyday problems of navigators, farmers, governors, and priests were left unresolved. Aristotle's great cosmology may have made men feel good, quieting doubts about the mechanism of the heavens, but it could not content the shipwrecked sailor, stranded for want of a reliable celestial chart. It could not console the farmer facing the failure of crops he had planted too late. Tax-collectors too, needed a calendar with which to plot the exaction of revenues. And priests were allowed no mistakes in timing their religious feasts and ceremonies: Midsummer's Day celebrations *had* to fall on Midsummer's Day! In all such practical matters the philosophical cosmologists, the explainers, were of little help.

Attempts to discern too quickly the super-system of the world, e.g., those of Aristotle and Aristarchus, came to be viewed as, if not beyond human powers, of no utility.

Mathematicians had probably then already to a great extent given up the hope of finding the physically true system of the world and had decided to look for a mathematical theory which would make it possible to construct tables of the motions of the planets. [19]

[17] Cf. Neugebauer, *Journal of Near-Eastern Studies* **4** (1945).
[18] "That this change occurred about the middle of the third century was a circumstance not unconnected with the simultaneous rise of the school of Stoic philosophy, which may be considered as a natural reaction against the idealism of Plato and the dogmatic systematizing of Aristotle. Both in abstract philosophy and in science the wish to get on more solid ground now became universal, and no science benefitted more by this realistic tendency than astronomy."                    (Dreyer, *op. cit.*, p. 151.)
[19] Dreyer, *op. cit.*, p. 157.

Young Greek mathematicians grew up with a choice (familiar enough today) between abstract thought about 'fundamentals', as against the sifting of reams of unsystematized, but accurately-recorded data. They were mostly drawn to the latter alternative; just as now young physicists find the 'successive approximation' empirical approach the more profitable. They pursued the narrow safe goals of accurate description and prediction, forsaking the ideal discerned by the Golden Age philosophers – comprehensive, cosmological explanation.

"Let nature *be* fundamentally unintelligible", they might well have argued; "Search only for the generalizations buried in stellar data-collections. Discern the geometrical plot which works. That is reward enough". This shift in attitude is a major change between the Golden Age and 'Hellenism'. Apollonios, Hipparchus, and Claudius Ptolemy advanced most ingenious mathematico-astronomical theories. These rivalled in complexity, and in requisite skill, the best that any prior geometer could have mustered. Historians sometimes view the transition from the Attic to the Hellenistic period, and thence to the Age of the Antonines, as one of unrelieved decline – so far as creative activity and intellectual originality are concerned. This is just not true within physics, astronomy, and mathematics – as the names of Euclid, Archimedes, Apollonios, Hipparchus, and Ptolemy are quite sufficient to establish. These men were not unravelling all the riddles of the universe simultaneously *a la* Plato and Aristotle. Like Eudoxos, they took on only those problems which were tractable enough to be wrestled one at a time. If creative, artistic, poetic, and speculative thought waned after Aristotle (which is also doubtful), scientific thinking of the most original, profound and uncompromising variety waxed stronger than ever. *Its* zenith was not in Aristotle, but in Ptolemy.[20]

---

[20]  Compare Bertrand Russell:
> "The mathematicians and men of science ... connected with Alexandria in the Third Century before Christ were as able as any of the Greeks of the previous centuries, and did work of equal importance. But they were not, like their predecessors, men who took all learning for their province, and propounded universal philosophies; they were specialists in the modern sense. Euclid, Aristarchus, Archimedes, and Apollonios, were content to be mathematicians; in philosophy they did not aspire to originality."
> *History of Western Philosophy* (London, 1946), p. 246.

*Pre-Ptolemaic Anticipations*

> The epicycle-on-deferent technique of Hipparchus and Apollonios
> Ptolemaic refinements

*Explanation and Prediction Again*

> Cosmology and astronomy
> Kinematics and dynamics

Astronomers sought now accurately to represent the movements of the planets *as observed*. They sought mathematical means of linking observations such that, when extrapolated into the future, predictions might become possible. The combinations of circular-spherical motions harnessed for this purpose were regarded only as calculi–devices for computing planetary positions at any time. The physical truth of such calculi (and their constituent geometrical devices) was never insisted upon, or even aspired for.[21]

In Book III of his mighty tract, *The Almagest*[22] Claudius Ptolemy credits Apollonios with knowing how an epicycle moving on a deferent can account for the apparent retrograde motions of the planets:

Apollonios, among others, examined one of the two inequalities – the one depending on position relative to the sun. Represented by means of the epicyclic theory, the epicycle is made to move in longitude ... etc. (cont'd below).

This passage contains the kernel of Ptolemaic astronomy, the dominant conceptual framework for understanding the cosmos, until the late 16th century.

What follows has two objectives. The first is to describe more graphically than has yet been done the elegance of the Apollonian-Hipparchean-Ptolemaic technique of epicycle-on-deferent. The second is to expose an implicit contention of several historians – that the inadequacy of Ptolemaic astronomy is somehow connected with its formally weak computational equipment.[23]

---

[21] Thus Geminus:

"For the *hypothesis* underlying the whole of astronomy is that the sun, the moon, and the five planets move at uniform speeds in circles, and in a sense contrary to that of the motion of the universe."

*Elements of Astronomy*, I (Trans. T. L. Heath, *Greek Astronomy*).

[22] *Al Majisti* in Arabic: 'The Greatest'.

[23] So prevalent is this view that as late as 1880 De Morgan wrote:

When compared with the calculational devices of our century, epi-cycle-on-deferent astronomy apparently comes out as a laughably primi-tive attempt to predict planetary perturbations. But to reason thus is fallacious.

First then, we will attempt a graphic account of the flexible beauty of the epicycle-on-deferent technique. As Ptolemy himself said (to con-tinue the earlier quotation):

... the epicycle is made to move in longitude in the order of the zodiacal signs (i.e., from west to east) in a circle concentric with the Zodiac, while the planet moves on the epicycle at a velocity which is the same as that of the anomaly; on that part of the epicycle farthest from the earth the motion is direct.[24]

Ptolemy here gives that example of epicyclical motion which provides the staple illustration in history of science textbooks (see Figure 27).

Chalcidius said of Herakleides that he envisaged such a motion for Mercury and Venus: we referred to this earlier.[25] Vitruvius amplified this.[26]

Martianus Capella was the most explicit:

Venus and Mercury ... place the center of their orbits in the sun; so that they sometimes move above it and sometimes below it [i.e., nearer the earth] ... the circles of this star [Mercury] and Venus are epicycles. That is to say, they do not include the round earth within their own orbit, but are carried around it latterly, as it were.[27]

As textbooks show, by reversing the direction in which the planet turns on the epicycle, an elliptical orbit results.

---

"On this theory of epicycles ... the common notion is that it was a cumbrous and useless apparatus, thrown away by the moderns".
   (*Dictionary of Greek and Roman Biography*, compiled by Sir William Smith [London, 1880], s.v. Claudius Ptolemaeus, p. 576.)
Dampier does not correct 'the common notion' clearly enough. He writes:
   "Its [the theory of Hipparchus and Ptolemy] one fault *from the geometric point of view* was the complications of cycles and epicycles it involved".
   (*A History of Science*, 4th ed. [Cambridge], p. 109, my italics.)
Even Abetti writes misleadingly in this connection:
   "It [the Copernican system] has the weakness of the epicycles, which could not explain the variable direction of the planet, due to its elliptic motion around the sun..." (*The History of Astronomy* [New York, 1952], p. 80.)

[24] Claudius Ptolemy, *Syntaxis mathematica*, ed. Heiberg (Leipzig, 1898–1903), III, 3.
[25] *Commentary on Plato's Timaeus*, 109.
[26] *On Architecture*, IX, 1.6.
[27] *On the Marriage of Philology and Mercury*, VIII, 859, 879, my italics.

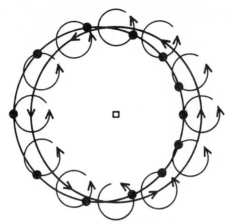

Fig. 27. Epicyclical motion.

This same effect could have been obtained without the epicycle – by letting the deferential circle's center □ itself move in a circle just the size of the epicycle above. As Ptolemy puts it:

The center of the excentric revolves ... whilst the planet moves on the excentric in the opposite direction ....[28]

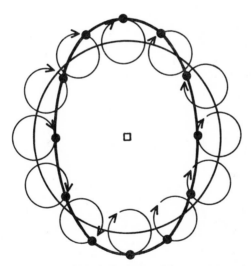

Fig. 28. Generation of elliptical orbit by circular motions.

[28] *Syntaxis mathematica*, III, 3.

Indeed, Ptolemy superimposes both demonstrations to reveal their perfect equivalence:

According to either hypothesis it will appear possible for the planets seemingly to pass, in equal periods of time, through unequal arcs of the ecliptic circle which is concentric with the cosmos ...[29]. It must be understood that all the appearances can be accounted for interchangeably according to either hypothesis, when the same ratios are involved in each. In short, the hypotheses are interchangeable.[30]

The epicyclical version of this technique of combining circular motions came most to be used in semi-popular expositions and texts. It was psychologically simpler, and it illustrated the stops and reversals of the planets more graphically than could be done by the excentrics. (Figures 28 and 29.) Besides, the latter technique could be employed *only* with Mars, Jupiter, and Saturn; Mercury and Venus required epicycles. By the uniform adoption of epicyclic technique then, the idea took root that all this was a *scientific system*. But that is a mistake, as will be argued in the sequel. Actually, in computational astronomy, the excentric tech-

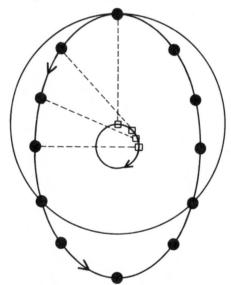

Fig. 29.    Geometrically-equivalent orbit generated eccentrically.

[29] Ptolemy understands the epicycle's explanatory potential better than Abetti, referred to earlier (cf. Note 23).
[30] *Ibid.*, and compare Book IX.

nique became more dominant. Since an immobile earth, once fixed, did not require reiteration by the working astronomer, he naturally chose that representation unencumbered by a circus of circles; it thus became somewhat easier for him to address 'the problem of the empty focus', to be discussed later.

Textbook expositions usually reach their zenith with Figure 30.

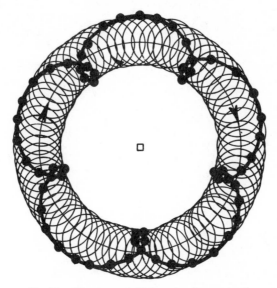

Fig. 30.   Circular cycloidal motion from epicycle.

With only this much of the epicycle-on-deferent story told, however, it would indeed appear that the Ancient's astronomical tasks were Sisyphean. The complex 'circumterrestrial' trajectories of comets, and the erratic wanings and wobblings of stars like Algol[31] – not to mention the variation-in-latitude of Mars and Venus – these seem to be motions which *in principle* elude a primitive technique such as described in Figures 27–30; it seems to be little more sophisticated than the theory of Eudoxos. But this is wrong. It must never appear that there might be dynamically significant motions which could elude Ptolemy's methods.

For example, by making the length of the epicycle's radius approxi-

[31] An eclipsing Binary in Perseus, the combined naked-eye magnitude of which ranges between 2.3 and 3.4.

mate to that of the deferent one attains perfect rectilinear motion along a diameter of the deferent (Figure 31).[32]

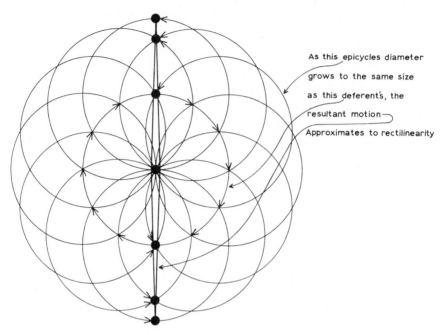

As this epicycles diameter grows to the same size as this deferent's, the resultant motion— Approximates to rectilinearity

Fig. 31.   Rectilinear trajectory from epicycularity.

[32] Compare Copernicus, *De revolutionibus orbium coelestium* (Thoruni, 1873), III, 4:
        "someone will ask how the regularity of these librations is to be understood, since it was said in the beginning that the celestial movement was regular, or composed of regular and circular movements ..."; "movement along a straight line is compounded of two circular movements which compete with one another ... [all] reciprocal and irregular movement is composed of regular movements ..." (pp. 165–166).
    Cf. also, C. B. Boyer, "Note on Epicycles and the Ellipse from Copernicus to Lahire," *Isis*, 1947, 38, 54–56.
    Historically, this motion is significant. Kepler makes Mars librate rectilinearly across its epicycle. (Cf. *De Motibus Stellae Martis*, IV, Ch. 58, in *Gesammelte Werke* (München, 1937).) Kepler realized that such a librational hypothesis could be equivalent to the elliptical orbit hypothesis independently discovered for calculating Mars' longitudes. (*Ibid.*, Ch. 59. Kepler does not actually undertake the construction just set out; but inasmuch as his librations are always 'librati in diametro epicycli,' Kepler's 'justification' of this rectilinear motion would have to be precisely what has been drawn above.) Cf. Book Three, Part I of this work.

By altering the velocity of a planet's east-to-west motion on an epicycle travelling west-to-east, one can approximate to a triangular figure
(Figure 32).

Indeed, one can take this to its rectilinear limit. (Figure 33.)

Fig. 32.   Near-triangularity from epicycularity.

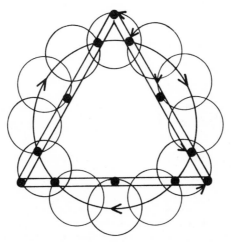

Fig. 33.   A rectilinear triangle from epicycularity.

When confronted with this triangular 'orbit', and the square one (Figure 34), several historians and philosophers were incredulous. Before having actually seen these figures, they thought them non-constructible by epicycle and deferent alone. Observers had to be assured that no trick, no 'juggling' of the epicycle speeds, had conjured these sharp-cornered figures.

Note, however, that both 'orbits' above are described with a rather broad line. This is theoretically significant. But since our second objective lies in the explanation of this broadening, let us return to the matter in a moment.

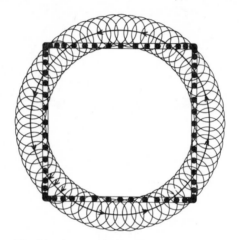

Fig. 34.   A square 'orbit' from epicycularity.

Consider first a few more 'orbital' possibilities contained within the ancient technique. By letting a second epicycle ride along on the first, a complex variety of ellipses becomes constructible (Figure 35).

Indeed, by varying the revolutional speeds of the secondary epicycles, an infinitude of bilaterally-symmetrical curves can be produced.[33] (Figure 36.)

[33] Even Kepler's intractable oviform curve, to determine the equations of which he implored the help of the world's geometers (Cf. *De motibus stellae Martis*, IV, Ch. 47, p. 297 "... appello Geometras eorumque opem imploro.") can be approximated quite closely with a third epicycle. (See Figure 37.)

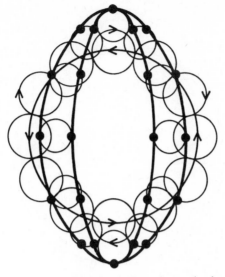

Fig. 35.   An infinitude of ellipses from epicycles.

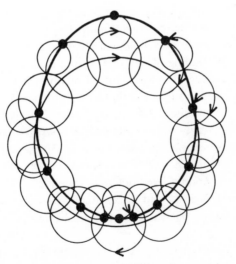

Fig. 36.   An infinitude of bi-lateral symmetries.

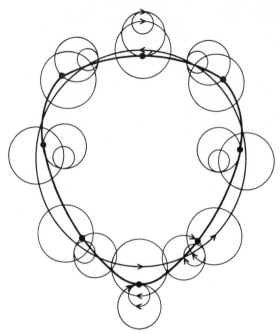

Fig. 37.   Kepler's 'Oviform Orbit' from epicycularity.

Some extraordinarily intricate, periodically-repetitive configurations are also at once constructible. (See Figure 38.)

In fact, an infinitude of *non-periodic* 'orbits' can also be generated. Simply add to the bundle of epicycles riding on a given deferent, and vary (at will) the speeds and directions of revolution for the component epicycles. Clearly, the range of complexity for orbits resulting from the epicycle device is unlimited. *There is no bilaterally-symmetrical, nor excentrically-periodic curve used in any branch of astro-physics or observational astronomy today which could not be smoothly plotted as the resultant motion of a point turning within a constellation of epicycles (finite in number)* revolving upon a fixed deferent. (See Figure 39.)

Bilaterally-symmetrical, and excentrically-periodic orbits that are curvilinear – these are one thing; but *rectilinear* polygons are quite another.

Return now to the triangular and the square 'orbits' depicted above (Figures 33 and 34). No one will doubt the formal power of an astronomical-geometrical technique which could permit a planet to move along even a square path if need be. We rarely put so heavy a demand on our own contemporary techniques – though we could do so if the heavens required it. The point is that Apollonios, Hipparchus, and Ptolemy could have done the same thing, had the heavens (as they observed them), required it. For all practical astronomy (in the 20th century as well as in the 2nd), one could, with a finite number of epicycles, generate a square orbit.

More precisely, if the square is a *visible* square – i.e., if its lines have some detectable breadth as well as length – then that square can be the resultant orbit of some *finite* combination of epicycles. Choose some arbitrarily small number ε (as small as you please, but greater than zero); it is possible to get the sinusoidal 'wiggle' (typical of epicyclical constructions) *so small* that its amplitude, i.e., the distance between its crests and troughs, will be smaller than ε. All this with a finite number of epicycles! Theoretically, of course, there must always be some slight cusp at the corners of the square so long as the number of epicycles remains finite. The cusps disappear, theoretically, only when the number of constituent

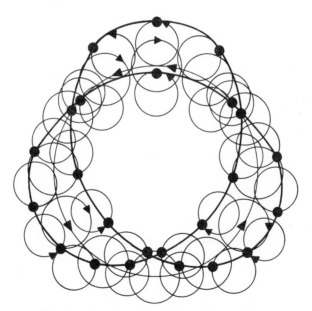

Fig. 38.   Complex, periodically-repetitive orbits.

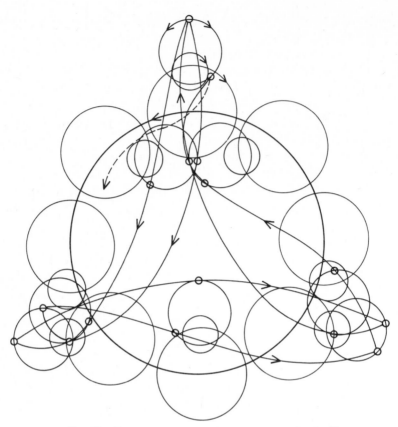

Fig. 39.   From symmetry to non-symmetry *via* epicycles.[34]

epicycles has gone to infinity. But in that case ε will have gone to zero
too; the square in question would be composed of *Euclidian* straight
lines (i.e., one-dimensional curves) – in which case the square would no
longer be visible anyhow. Thus, no matter how finely the square is drawn,
if it can be *seen*, then there is some finite number of epicycles of whose
resultant motion the square is the construct.[35]

---

[34] The bi-lateral symmetry of the solid-line orbit, gives way to the non-periodicity of the
dotted-line orbit when the planet's motion (on the 'external' epicycle) is reversed from
E→W at 2 turns per deferential revolution to W→E at 2–2/3 turns per deferential revolution.
[35] Cf. the Appendix.

# APPENDIX

(A) Let us represent points in the plane by complex numbers $z = x + iy = re^{i\theta}$; the corresponding Cartesian coordinates will be $(x, y)$, and the corresponding polar coordinates will be $(r, \theta)$.

Remember that the addition of two complex numbers

$$(z, z') \to z + z',$$

corresponds to vector addition of the position vectors of the corresponding points. Thus:

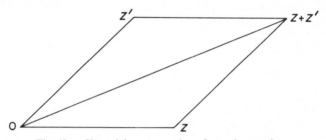

Fig. 40a. Vectorial representation of complex numbers.

(B) A uniform circular motion with center $c$, radius $\varrho$, and period $T$, may be represented by:

$$z = c + \varrho e^{(2\pi it/T) + i\alpha},$$

where $t$ denotes time, and $\alpha$ represents the initial phase of the point.

(C) Suppose now that a point $A$ is moving in the way described by the equation:

$$z = f(t).$$

Suppose also that $B$ is moving relative to $A$ in a circle of radius $\varrho$, with period $T$ and initial phase $\alpha$; then the motion of $B$ is given by the equation:

$$z = f(t) + \varrho e^{(2\pi it/T) + i\alpha}.$$

We can then think of $B$ as moving on an epicycle carried by $A$.

(D) We see immediately that the superposition of a new epicycle (one now carried by $B$) is equivalent to the addition of a new term:

$$\rho e^{(2\pi i t/T) + i\alpha}.$$

This is added to the expression for $z$. This term can also be written:

$$\rho e^{i\alpha} e^{(2\pi i t/T)},$$

or, more briefly,

$$a e^{ikt},$$

where $a$ is a non-zero complex number, and $k$ is real.

(E) Note that *any* form of retrograde motion corresponds simply to taking $T$ (or $k$) negative.

(F) A motion given by the superposition of $n$ epicycles would be given by an equation of this form:

$$z = a_1 e^{ik_1 t} + a_2 e^{ik_2 t} + \cdots + a_n e^{ik_n t}.$$

(G) Now suppose that we are just *given* a periodic motion in the plane by the equation:

$$z = f(t).$$

We may assume now that the period of motion is $2\pi$ (even if it becomes necessary to change the time scale somewhat).

Suppose that $f(t)$ is a sufficiently well-behaved function [it will be enough that $f(t)$ is continuous and of bounded variation – this is a natural enough condition to impose; that $f(t)$ is taken as continuous and of bounded variation means only that the resultant orbit is a continuous curve of finite length.] Then it is well known that we can write:

$$f(t) = c_0 + (c_1 e^{it} + c_{-1} e^{-it}) + (c_2 e^{2it} + c_{-2} e^{-2it}) + \cdots =$$
$$= \sum_{n=-\infty}^{\infty} c_n e^{int},$$

this series being *uniformly* convergent.

Let us now write:

$$S_N(t) = \sum_{n=-N}^{N} c_n e^{int}.$$

Choose now some small number $\varepsilon > 0$. Since the series is uniformly convergent, we can now find $N_0$ such that

$$|f(t) - S_N(t)| < \varepsilon,$$

for all $N \geqslant N_0$ and all $t$. In other words, if we consider the two orbits:

$$z = f(t), \qquad z = S_N(t),$$

the distance between corresponding points of these two orbits remains *less than* $\varepsilon$ for all time.

Thus the original orbit, $z = f(t)$, can be replaced, with as small a loss in accuracy as we please, by an orbit $z = S_N(t)$; that is, some finite superposition of epicycles.

(H) The function defining a square or a triangular 'orbit' satisfies these conditions completely; the above considerations therefore apply without qualification to these special cases illustrated earlier. (Figures 33 and 34.)

(I) The superpositions we have used so far are special ones; for the periods of the epicycles are:

$$\pm 2\pi, \ \pm \pi, \ \pm \tfrac{2}{3}\pi, \ \pm \tfrac{1}{2}\pi, \ \pm \tfrac{2}{5}\pi \ldots.$$

In particular, they are commensurable with one another.

(J) However, it must be stressed here that even non-periodic orbits can be represented by such a superposition of epicycles, provided only that we allow incommensurable periods. The basic theorem would be:

Let $z = f(t)$ be a complex-valued function of the real variable $t$, and suppose that $f(t)$ is a uniformly almost-periodic function of $t$. (See H. Bohr, *Fastperiodische Funktionen* (1932) for all the necessary definitions); then we can approximate to this motion as closely as we please for *all time* by an expression of the form:

$$z = a_1 e^{i\lambda_1 t} + a_2 e^{i\lambda_2 t} + \cdots + a_n e^{i\lambda_n t},$$

i.e., a superposition of epicycles of radii $|a_1| \ldots |a_n|$, and periods $2\pi/\lambda_1 \ldots 2\pi/\lambda_n$.

Further observations are now in order.

All of our diagrams have been presented in the plane. (How else in a book?) This is a 'technical' limitation, however, and should not obscure the fact that geocentric astronomy was designed as much to cope with a planet's aberrations in latitude as with those in longitude. By tilting the epicycle's axis of rotation, further variations normal to the plane are achieved, as we shall see. By extension of the demonstrations above, it proves possible to move a point (within a finite framework of epicycles) over the surface of a *cube*, a *pyramid*, an *ovoid* – or indeed, over any solid figure generable *via* the foregoing diagrams construed as sections. This is inherent in the ancient technique.

Formally, epicycle theory is not a 'closed book' – a musty collection of ancient thought-recipes. The famous 'brachistochrone' problem, introduced by Johann Bernoulli [36] has deep affinities with epicyclical motion considered as under the influence of gravity alone. This problem, along with the so-called 'isoperimetric' problem, led historically to the development of the calculus of variations by Euler and LaGrange, and ultimately (through Maupertuis, Fermat and Hamilton) into the general theory of variational principles – a powerful formal tool in mechanics, optics, electrodynamics, and even in relativity and quantum theorry. Modern physicists often face dynamical problems analogous to the very difficulties which the intrepid Ptolemy faced. Indeed, some of his problems require careful analysis for their full solution even today.

Lunar motions, as well as those of the satellites of Mars, Jupiter and the other planets, are almost purely epicyclical. Mars' moons, moving in opposite directions around Mars, generate a cluster of difficulties within epicyclical theory especially when complicated by perturbational considerations. Our moon describes what is basically the path of an epicycle moving on a deferent (the earth's orbit) slightly excentric to the

[36] *Acta eruditorum*, 1696.

sun. Moreover, subtle variations in the ellipsoidal character of the earth's orbit, as well as in the moon's own librations and perturbations, stretch epicycle theory to its utmost as a contemporary astronomical tool. Had he possessed the data *we* have, Ptolemy would have been quite at home with some of our problems in lunar dynamics.

Ptolemy's mathematics was then, as powerful – for the special problems before him then – as our own in dealing with the same perplexities. It is untrue then, that the epicycle technique restricted ancient astronomical thought.

Apollonios himself, if Ptolemy is correct, could have constructed most of the celestial motions known to us: this, only with his circles, spheres and epicyclical ingenuity.

Why did Hellenistic astronomers, like their Attic predecessors, insist on the principle of circularity, however? Archimedes had explored the properties of the ellipse – and paraboloid curves were familiar enough. Epicyclical combinations produced wholly non-circular orbits (cf. Figures 28–39). If prediction was the major criterion of success in astronomical theory, why not employ non-circular curves initially and drop this philosophical facade of circles and circularity? Formally speaking, all curves are equally fundamental: any one of them can be generated from any other one, given only a suitably-defined transformation space. Newton himself often treated the circle as being merely a degenerate ellipse.[37] Why, then, this studied protocol of building up the heavens out of spheres, discs and rings? Why these annular mannerisms?

The answer: the ancient nature-philosopher's need for total explanation (and cosmological speculation) was not wholly dead in the 'predicters', Apollonios, Hipparchus and Ptolemy. Explanatory speculations were ruled out of their explicitly formal astronomies. Running parallel with mathematical developments in astronomy, cosmological inquiries continued, albeit subdued. Often, they were extra-curricular asides by the very men who had forced them out of computational astronomy; today, physicists who proclaim their 'serious' calculations free from 'mere philosophy and methodology' often cram their Prefaces, In-

---

[37] Cf. *infra* Book Three. Part II. *Editor's note:* This part of the work was incomplete at the time of Hanson's death.

troductions and Conclusions with the merest philosophy and the least examined methodological trivia.[38]

Astronomers (including Ptolemy) felt that *if* a physically true picture of the universe was to be had, it would consist in some modification of the nested-sphere cosmos imagined by Aristotle. The Stoics were also of this opinion. Being *the* influential writers of the Hellenistic period, their 'Aristotelian notions' were much publicized.[39]

Hence, at least one physical principle had to be common both to cosmology and to astronomy. *Ultimately All Celestial Motions are Circular in Nature* – a reification of the First Fact of The Heavens as encountered in Part I. Anything else was unthinkable. Thus:

> ... it is ... the aim of the mathematician to show forth all the appearances of the heavens as products of regular and circular motions ... [and to] separate out the particular regular motions from the anomaly which seems to result from the hypotheses of circles, and show forth their apparent movements as a combination and union of all together ...[40]

It is

> ... necessary to assume in general that the motions of the planets in the direction contrary to the movement of the heavens are all regular and circular by nature, like the movement of the universe in the other direction. Their apparent irregularities result from the positions and arrangements of the circles on their spheres through which they produce these movements, ...[41]

Ptolemy himself remarks that it is of the nature of 'the incorruptible' to move in circular fashion, from which it follows for him that the proper motion for all celestial bodies was perfectly circular. Indeed, there is more than one example in ancient astronomical writings of the following inference:

> ... but hypothesis H would entail that the planet move in a non-circular way; so it follows that not-H.[42]

---

[38] Even Newton's *Scholia* are open to this charge; they do not logically encapsulate his inferential procedures in *The Principia*, however much they may purport to do so.
[39] Cicero, Seneca, and the doxographic writers, were unfamiliar with the mathematical theories of Apollonios. Yet they adopted wholesale the cosmological opinions of the Stoics, which were identical with what had been written down by Aristotle, or his students.
[40] Ptolemy, *Almagest*, III, 1.
[41] *Ibid.*, III, 3.
[42] This argument once gripped young Kepler, cf. Book Three, Part I.

In short, it had become part of the concept *planet* that any object, to qualify *must* move circularly.

So even a predicting device, which was what astronomy was for Apollonios, Hipparchus, and Ptolemy – were it to make contact with the same heavens discussed by cosmologists – had to share with the latter this fundamental and characteristic property, perfect circularity.

Thus the interim figure, Theon (of Smyrna), seeks the best of both worlds.[43] He conceives of Apollonios' epicycles (qua geometrical calculations) as *actually existing solid spheres*. This was a *via media* between Peripatetic cosmology, and epicyclical astronomy; between Aristotle's urge to explain and Apollonios' wish to predict the planetary perturbations. The most authoritative cosmology (Aristotelian), tended in this way to support the most accurate astronomy (epicyclical) against all rival theories, and *vice versa*.[44]

The stage was now set for Hipparchus of Rhodes, the greatest original astronomer of all antiquity. Besides founding trigonometry, Hipparchus discovered the precession of the equinoxes – the advance of the point where the Ecliptic and the Celestial Equator intersect. Theon even credits Hipparchus with the invention of the theory of epicycles.[45] Indeed, all of Ptolemy's *Syntaxis Mathematica* may be thought of as a collection, a completion and a detailed development of Hipparchus' work, notwithstanding the brilliant contributions of Ptolemy himself.

Having before him a vast store of Babylonian and Alexandrian observations, to which he added many of his own, Hipparchus sought a possible orbit for the sun around the earth – one which would account for the differing durations of the seasons. Like Apollonios before him (and Ptolemy later) he offered two equivalent hypotheses, either one of which accorded with the observations – to within 1° of celestial arc.

Either device permitted accurate predictions of the sun's future positions. Thus, adopting the Eccentric mapping in Figure 41, let *C* be the center of the circle which the sun describes about the earth. Let

---

[43] *Theonis Smyrnaei Liber de Astronomia*, Ed. Martin, Paris, 1849, p. 282.
[44] Adrastus, a much later Peripatetic, still redoubtably supports the notion of solid celestial spheres.
[45] *Op. cit.*, p. 300.

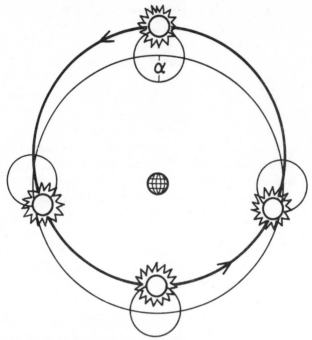

*Epicyclic:* during the tropical year the sun moves on a epicycle (radius α), whose center moves east around the earth.

Fig. 40b.   Hipparchus' epicyclical account of the seasons.

*V* and *A* be the vernal and autumnal equinoxes respectively – where day and night on earth are of equal length; *S* and *W* are the summer and winter solstices – where the celestial equator and the ecliptic are farthest apart. Given the length of the seasons (as actually observed), Hipparchus computed the eccentricity ($e$) to be $\frac{1}{24}$ (i.e., of the radius of the circle), and the arc to apogee as being 65°30'; he assumed that these figures were constants. Here is a typical use of the eccentric circle so dominant in technical astronomy before Kepler. It has *the problem of the empty focus* built into it. Given the fact known to us (but not the Greeks), that planetary orbits are ellipses – but given also the 'rule' of the ancient astronomy, viz., that *all calculations are to be carried out by using only circles and spheres* – the standard challenge was then always to locate the eccentricity of a chosen planet's (circular) orbit, so that it

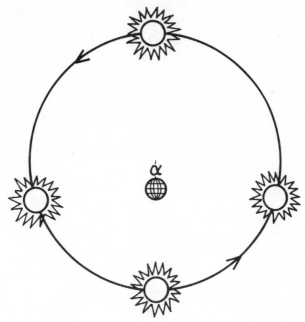

*Eccentric:* during a year the sun moves east in a circle, from whose center the earth is
separated by α.

Fig. 41.   Hipparchus' eccentric account of the seasons[46].

approximates to being just half the distance between the two foci in the
*actual* orbit (see Figure 43).

The problem of Hellenistic astronomy was always: how far from the
deferential circle-center should earth be placed? Knowing (as we do)
the actual orbit, this can be restated: how far should the circle's center be
moved towards an elliptical focus – and how far *away* from the other
(empty) focus? This could be called 'the deferential calculus' of ancient
astronomy.

---

[46] Compare Geminus:

> "... The sun revolves at a lower level than the signs (i.e., is closer to the
> earth), and moves on an eccentric circle ... For the sun's circle and the
> zodiac's circle have not the same center, but the sun's course is divided
> into four unequal parts ... The sun, then, moves at uniform speed through-
> out, but because of the eccentricity of the sun's circle, it traverses the qua-
> drant of the zodiac in unequal times."
>
> (*Elements of Astronomy* I (Trans. Heath *Greek Astronomy*).)

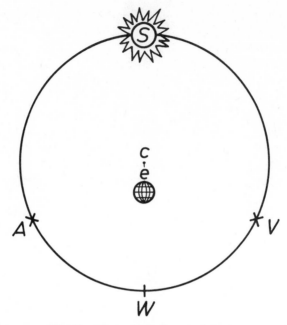

Fig. 42.   Hipparchus' solar calculations.

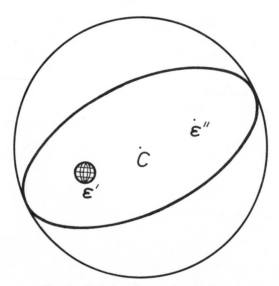

Fig. 43.   Relation of eccentric and ellipse.

The moon was difficult to cope with in this respect, but Hipparchus still managed to deal with his first inequality by an eccentric or an epicycle.[47] But that some other additional 'inequality' was involved in lunar motion soon became clear. Hipparchus could cope with the new or full moon, but the first and last quarters eluded his theories altogether. Similarly, individual planets proved particularly recalcitrant.

As with Apollonios, Hipparchus was primarily a mathematician. However, his philosophical–cosmological ideas run parallel to his technical astronomy. Thus, although he demonstrated that the two are formally equivalent, Hipparchus always prefers epicycles to movable eccentrics[48]; it seemed more reasonable that celestial bodies should be ordered symmetrically 'round the center of the universe. Theon continues

... though [Hipparchus] was not a natural scientist and did not detect which was the true and natural motion of the planets (and which accidental and merely apparent), nonetheless he thought the epicycle of each to move on a concentric circle – the planet being on the epicycles' circumference.[49]

This much reveals Hipparchus as a mathematician concerned not so much with the physical understanding of celestial phenomena, as with the accurate description and prediction of their occurrences.

Ptolemy, however, puts the matter differently from Theon:

... because he had not himself been given as many accurate observations as he has passed on to us ... [Hipparchus] only examined the sun and the moon, demonstrating that their revolutions could be perfectly accounted for by combinations of circular and uniform motions. As for the five planets however ... he has not even begun to theorize, contenting himself only with the systematic gathering of observations and the demonstration of how they refuted the mathematical hypotheses of his day.[50]

Ptolemy continues:

Indeed, [Hipparchus] explained not only that each planet has two types of inequality, but also that their retrogradations are variable; other geometers had shown but one inequality, and one arc of retrograde motion. Hipparchus did not believe that these events could be

---

[47] The 'first lunar inequality' is simply the equation of its center – a description actually required by the *de facto* elliptical form of the moon's orbit.
[48] Cf. Theon (Martin), p. 300.
[49] *Loc. cit.*
[50] *Almagest*, IX, II. Note this as an almost perfect exposition of the 'Hypothetico–Deductive method', 2000 years before Mill, Carnap, Popper, Reichenbach, Hempel and Braithwaite.

represented either by eccentrics or epicycles, but that … the two hypotheses would have to be combined … .

It seems, therefore, not that Hipparchus regarded explanatory physics as less worthy than a mathematical computer for planetary positions, but rather that complex celestial phenomena forced him to abandon hope of any complete planetary theory. Hipparchus' predecessors had sought only to account for the annual aberrations of a planet when in opposition.[51] They were unaware that, over the years, the arc of retrograde motion varies in length.[52] To detect this variation requires long and careful observation; e.g., Saturn requires 29458 *years* to traverse the stellar sphere once. Hipparchus felt (quite properly) that no theory which failed to account for these variable retrograde arcs could be adequate. Thus he elected to gather observations which, when added to those of the three intervening centuries, provided Ptolemy with the data required for his own planetary researches.

The most that was known to Hipparchus, therefore, was that the planets, when opposite the sun (seen from earth), appeared to halt and retreat before continuing ahead. Even the moon seemed to depend in some such way upon the sun; Hipparchus discovered its velocity at quadrature to be variable. Every wandering star seemed connected with the sun: every ancient astronomical problem concerned a *sun-dependent anomaly*.[53] But this suggested to almost no one that the universe might be other than geocentric and geostatic. Astronomy had become a powerful instrument for calculation. But even its disclosure that all sub-stellar celestial phenomena seemed to be functionally connected with the sun could not weaken the cosmological commitments running so deep in the intellectual stream of the period. The sun continued to be treated as but one more circumterrestrial object – somewhat as if a traffic policeman were regarded as being but one more pedestrian, despite his correlative connections with all motions around him. In the 16th century this functional dependence was at last recognized as basic to the planetary traffic system.

When an ancient astronomer sought to *predict*, he used formal devices

---

[51]  Cf. Figure 17.

[52]  This is the effect of the elliptical form of 'out-of-phase' planetary orbits, the changes in arc-length and apparent velocity resulting entirely from the shift in terrestrial view-point.

[53]  Compare Ptolemy, *Almagest*, Book IX, p. 271, Note 1.

which by themselves could not help one to understand the phenomena being forecast. When asked to *explain* a celestial event he usually 'changed the subject' and appealed to a cosmological picture depicting every thing at once, but which could predict nothing in particular.

Claudius Ptolemy's *Almagest* perfectly illustrates all these things. Astronomy had progressed too little in the 260 years since Hipparchus.[54] Certainly the theory of epicycles and eccentrics had not advanced one turn beyond its beginnings in Hipparchus and Apollonios. The task Ptolemy set for himself was to gather, collate and summarize, all astronomical knowledge – much resembling what Euclid had done for geometry four centuries before.

Ptolemy starts[55a] with a survey of the presuppositions of all astronomy. "... the heavens are spherical and move circularly (around a fixed axis) ...". How else to explain the circumpolar star's motions in perfect circles, and why other stars rise and set always at the same points on the horizon? "... The earth, in figure, is sensibly spherical ...". How else to explain the hull-first disappearance of departing ships and the peak-first re-appearance of mountains when sighted from sea? "(Earth)... lies right in the middle of the heavens, like a geometrical point (center)." If it did not, one side of the heavens would appear nearer, brighter, and larger than the other sides; the horizon would not bisect the celestial equator; the ecliptic would not be divided equally. "... whatever figure other than the circle be assumed for the movement of the heavens, there would have to be unequal linear distances from the earth to parts of the heavens... but this is not observed to happen... absolutely all the appearances contradict other opinions [as to the non-sphericity of the heavens]...". Besides, all heavy bodies 'fall to the center' of the heavens. Thus earth, which is the heaviest element [of the Four Elements] falls to the center. Fire, the lightest, falls away from the center to the circumference of the universe.

No stellar parallax is ever observed. This observational fact, which confounded heliocentrists for thousands of years (until Bradley and

---

[54] Ptolemy records but two observations as having been made in all that time; an occultation of the Pleiades in 92 A.D. (observed by Agrippa in Bithynia), and two occultations of Spica and $\beta$ Scorpii (observed by Menelaus at Rome in 98 A.D.).

[55a] *Almagest*, I.

Bessel), proves continually to be the strongest empirical support for the Ptolemaic geostatic formulation. Not until Bradley's discovery of stellar aberration (in 1729) – the discovery that slightly different angular positions of the fixed stars are observed, according to which direction the orbiting earth moves across their rays – and Bessel's discovery (in 1838–39) of an observable parallax in those very same fixed stars – not until then was the sting taken from this important geostatic argument.[55b]

Euclid had already anticipated much of Ptolemy:

... The universe is a circle in shape; for if it had been in the form either of a cylinder or of a cone, the stars taken on the oblique circles bisecting the equinoctial circle would, in their revolution, have seemed to describe not always equal semicircles, but sometimes a segment greater than a semicircle, and sometimes a segment less than a semicircle. For, if a cone or a cylinder be cut by a plane not parallel to the base, the section arising is a section of an acute-angled cone, which is like a shield (i.e., an ellipse). Now it is clear that, if such a figure be cut through its center lengthwise and breadthwise respectively, the segments respectively arising are dissimilar; it is also clear that, even if it be cut in oblique sections through the center, the segments formed are dissimilar in that case also; but this does not appear to happen in the case of the universe. For all these reasons, the universe must be circular in shape, and revolve uniformly about its axis, ...[56]

Aristotle stresses the empirical fact that:

... There is nothing in the observations to suggest that we are removed from the center [of the universe] by (even) half the diameter of the earth.[57]

This, of course, makes geocentrism just as difficult to maintain as any doctrine in which the earth moves through a small orbit; Heath and Schiaparelli both felt that for the Pythagoreans "parallax is as negligible in the one case as in the other."[58]

---

[55b] As Ptolemy observes:
> "... no change has hitherto taken place in their position [those of fixed stars] with respect to each other, but the configurations observed by Hipparchus are seen to be absolutely the same now. This is true not only of those inside the zodiac with respect to each other or those outside, ... but also of those inside with respect to those outside and far off."
>
> (*Almagest*, VII, 1).

[56] *Phaenomena*, Preface.

[57] *De Caelo*, II, 13.    *

[58] Cf. *Aristarchus of Samos*, pp. 100–101. Strictly speaking, absence of stellar parallax ought to perplex any theory in which the astronomer's eyeball is not *de facto* fixed at the center of the universe; something no ancient theory ever did maintain. The issue thus put becomes not one of philosophical principle, but merely of degree. If being 4000 miles from the Universe's absolute center generates no stellar parallax in the eye of the traveller – because that distance is too small – then the hypothesis of an earth travelling

Ptolemy also argues:

Besides, there *must* be some fixed point of reference.[59]

Common sense reveals that a pebble at rest on the floor of a chariot will fall behind when the chariot lurches away. Is it not equally clear that clouds, leaves, and even animals, would fall behind in space if the massive earth had to move fast enough to account for the diurnal motion of the fixed stars and the annual motion of the sun?[60]

This again is the voice of a strict observationalist:

and we shall try and show each of these things, using as beginnings and foundations for what we wish to find the evident and certain appearances from the observations of the ancients and our own, and applying the consequences of these conceptions by means of geometrical demonstrations ... [61]

Hipparchus' theory of the sun appears unchanged in the *Almagest*[62]; this is somewhat odd since Hipparchus' errors in solar position amount sometimes to almost 100′.[63]

But in lunar theory Ptolemy's originality discloses itself, as does also his attitude towards astronomy and cosmology, mathematical physics, prediction and explanation. Because Hipparchus' epicycle-on-deferent hypothesis left considerable errors when the moon was observed at syzygy[64] and also sometimes at quadrature. It appeared *prima facie* that the epicycle's size had to vary in length, expanding and contracting in circuit. But this was unthinkable. Ptolemy's singular innovation here consisted in suggesting that the lunar epicycle's center ($E$) moves east on an eccentric circle; its angular velocity, however, is not uniform with respect to the center ($C$), but rather with respect to the Earth.

In Figure 44, the line connecting the center and the apogee ($A$) of the eccentric circle itself rotates to the west, such that the angle $A$–earth–$E$

---

through an orbit 36 times greater (in diameter) than our distance from the moon should not be dismissed just because of an absence of stellar parallax. *That* may be too small a distance as well. So here, as elsewhere, the observations are never theoretically conclusive.

[59] Another singular argument!.

[60] *Almagest*, I, 7. (Cf. pp. 18–21 *supra*.)

[61] *Ibid*.

[62] Much as Eudoxos' work appears unchanged, but unacknowledged, in Euclid's *Elements*, Book V.

[63] Tannery, *Recherches*, pp. 169–171.

[64] Whenever earth, moon and sun lie along a straight line, whatever their order.

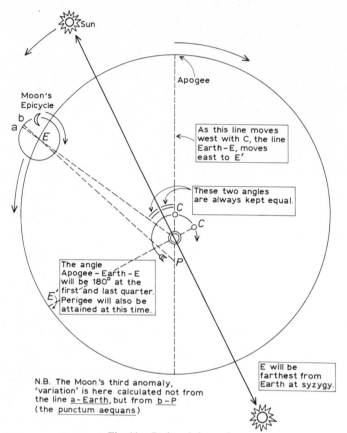

Fig. 44.  Ptolemy's lunar theory.

is always twice the moon's angular distance from the line connecting earth and sun.[65] Since the maximum that this lunar-solar angular distance can attain (either way) is 90° (= quadrature) the angle $A$–earth–$E$ must be 180° at the first and last quarter. Then $E$ will be farthest from the earth at syzygy. It will be nearest at quadrature, when the moon is at right angles to the line earth–sun. By this intricate and ingenious com-

[65] The angle $A$–earth–$E$ is equal to twice the elongation (angular distance) of the moon from the sun, being 180° at the first and last quarter $(E')$. $C$ (the deferent's center) therefore moves 'backwards' $(E \rightarrow W)$ with a velocity equal to twice the elongation minus the argument of latitude – so that $C$, in a synodic month describes a small 'clockwise' circle around the earth: this sets the timing for the rest of this lunar machine.

bination of epicycle and moving excentric Ptolemy accounted for the second inequality in the moon's motion; its hastening ahead of, or falling behind, the positions calculated even *after* finding the deferential eccentricity needed to cope with the first inequality – its deviation from perfectly uniform, circumterrestrial motion. This second anomaly we now call 'evection'. But a third error remained.

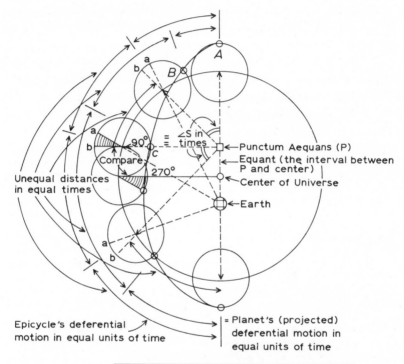

The de facto result of this innovation was to move the planet on its epicycle at inconstant velocities. For although presented as a planet in uniform circular motion on an epicycle whose deferential motion is variable - this comes close to saying that the epicycle's translation is uniform and invariant while the planet moves irregularly. Compare the planet at 90° on its epicycle with the epicycle at 270?

Fig. 45.   The 'Punctum Aequans'.

To correct this residual error (which we know as 'variation') Ptolemy calculates the anomaly not from the line *a–earth* (as had been done before), but rather from the line *b–P*. (Cf. Figures 44 and 45.)

This theoretical sidestep became an evasive lunge which ultimately tore the fabric of the Ptolemaic heavens. It was an arbitrary, an *ad hoc* supposition – invoked simply to cope with these particular observations.[66]

Much later on, astronomers began to doubt whether prediction *simpliciter* was enough to compensate for the use of such unprincipled procedures. Success in prediction certainly could not *justify* such procedures.[67] Students of the sky came to hope for some set of physically intelligible explanations of *why* such predictions did in fact succeed.[68]

Since no astronomical instrument in 200 A.D. was accurate to within more than 10′ (of celestial arc), Ptolemy's lunar theory determined the

---

[66] "Let no one, seeing the difficulty of our devices, find troublesome such hypotheses. For it is not proper to apply human things to divine things nor to get beliefs concerning such great things from such dissimilar examples ... But it is proper to try and fit, as far as possible, the simpler hypothesis to the movements in the heavens; and if this does not succeed, then (to try) *any hypothesis* (which is) *possible*. Once all the appearances are saved by the consequences of the hypothesis, why should it seem strange that such complications can come about in the movements of heavenly things? ... It is not proper to judge the simplicity of heavenly things by those which seem so with us, when here not even to all of us does the same things seem likewise simple ...."

(Ptolemy, *Almagest*, XIII, 2 (my italics).)

[67] Compare my book. *The Concept of the Positron*, Cambridge, 1963, Ch. II.

[68] In *De Revolutionibus Orbium Coelestium* (IV, 2) Copernicus sharply criticizes Ptolemy's high-handed manner of dealing with the third lunar anomaly. He notes a practical result of this contriving of movements: the moon can be considered to *move irregularly* about its own epicycle center. Copernicus argues that Ptolemy's initial violation of *the* principle of celestial mechanics (that an epicycle's movement on *its* deferent must always be uniform and unalterable), is what generates this further irregularity.

Ptolemy would have had a reply. Even if these *ad hoc* devices move planets on their epicycles irregularly – and the epicycles move inconstantly with respect to the centers of their deferents – these irregular movements can yet be derivatively computed, described and predicted by means of circles and regular movements alone; this is all that is necessary. Ptolemy's device may thus be not completely *ad hoc*. His technique of dealing with this third lunar anomaly does not require its own principles and computational devices *ab initio*. It is definitely connected with what Ptolemy might construe as *the* commitment of all astronomy, viz., that *every celestial computation can be compounded of perfect circles and their regular movements*. Perhaps Copernicus feels Ptolemy to have produced the non-circularities too soon in his calculations! Even Copernicus ends up with them sooner or later, when the Circularity Principle fails him – but invariably later than Ptolemy. Well, the Principle failed Ptolemy's lunar theory very early.

observed positions in a satisfactory manner. What more could one ask? Nothing observable could disconfirm such a calculation; if *apparentias salvare*, then [69] non-disconfirmation was an empirical argument in the calculation's favor. So, as a computing device, this Ptolemaic scheme was effective. Copernicus would not allow that it left 'other things equal', however: it ruined astronomy's link with rationality – this last requiring uniform, circular, undisturbed and perfect celestial motion.

Nor could Ptolemy begin to comprehend the true spatial positions of the moon relative to earth.[70] Indeed, according to his theory the moon's diameter at perigee ought to cover a whole degree![71]

Nowhere can Ptolemy conjecture how a planet's *actual* positions are related to his 'phenomenalistic' computing devices.* For any particular planet Ptolemy discovers some ratio between the radii of deferent and epicycle which permits him to plot the retrograde arc in close accordance with what is observed. Thus he comes to grips with our 'problem of the empty focus'. He can predict where familiar points of light – called 'Venus', 'Mars', 'Saturn', 'Jupiter' – will appear against the spherical background of fixed stars. But for this victory Ptolemy suffers a defeat concerning just how far from the earth, and from each other, the substellar bodies really are. By considering obscurations, and eclipses, Ptolemy can determine (as did his predecessors) which planets are nearest to us and which are farther away. But a precise quantitative account is utterly beyond him. His lunar theory is a case in point. To calculate the moon's celestial latitude and longitude (by epicycle and excentric), Ptolemy must grant a diametral variation of the lunar disc ranging from 31′30″ at apogee to 1° at perigee! This, again is quite contrary to the facts. Ptolemy's lunar object at perigee requires an appearance that would dwarf the most conspicuous 'harvest moons'; it would be twice the size of the sun at noon.

Like everyone else before and after him, Ptolemy *speculated* about

---

* Cf. Editor's note on p. 5 *supra*.

[69] Other things being equal (i.e., *aequalitatem tueri*).

[70] Compare *Almagest* IV, 5.

[71] His eccentric theory requires the moon's distance from earth to vary by as much as 34 to 65, nearly one to two. But, observation reveals that this is simply not true. (Compare Copernicus, *De Revolutionibus* ..., IV, 2.) The lunar diameter in fact varies as 55 to 65, just as an epicyclical theory *without* any additional eccentric allows. So here the Ptolemaic theory, in striving for *apparentias salvare vis-à-vis* the moon's angular translations, is unable to 'slow down' for our satellite's slight variations in diameter.

the relative ordering of the sun, moon and planets, and the distances between their several orbits.[72] But he thinks there is no way of proving which order is factually correct, since none of the planets have any sensible parallax:

There is no other way of getting at this [problem of planetary order] because of the absence of any sensible parallax in these stars, from which appearance alone linear distances are gotten, ... [We will use] the sun as a natural dividing line between those planets which can be any angular distance from the sun, and those which cannot but always move near it.[73]

Ptolemy's conception of the importance of parallax in determining celestial distances was sound: that is just how Bessel, Struve and Henderson determined the distances of 61 Cygni, Vega and α Centauri in 1840. And his recognition of the sun as pivotal in analyses of plane-tary ordering was also prophetic – for his 'sunbound' vs. 'sunfree' distinction corresponds to our 'inferior–superior' (or 'interior–exterior') contrast. Mercury and Venus were sunbound to within (app.) 25° and 45° respectively: the others were unfettered. His thoughts about order and distance were thus unconnected with Ptolemy's astronomical techniques. As he remarks in Book II of his *Hypotheses Concerning The Planets*,

the heavenly bodies suffer no influence from without; *they have no relation to each other*; the particular motions of each particular planet follow from the essence of that planet and are like the will and understanding in men.

Ptolemy now frankly concedes that epicycles and eccentrics merely pro-vide a means for calculating, describing and predicting the apparent places of sun, moon, and planets. They depict neither the *actual* physico-geometrical disposition of the world, nor do they hint at why certain combinations of circles result in successful predictions while others fail.

Perhaps Ptolemy paid too high a price for his moderately successful forecasting machine? The principle of uniform circular motion he violated – by inventing a point-other-than-the-center-of-the-deferent

---

[72] *Almagest*, IX, 1, II. Anaxagoras, the Pythagoreans, Plato, Eudoxos, and Aristotle placed the planets in this order: Moon, Sun, Venus, Mercury, Mars, Jupiter, Saturn. The later Stoics shifted the Sun from second back to fourth place. And Hipparchus, Geminus, Kleomedes, Pliny, Pseudo-Vitruvius, the Emperor Julian, and also Ptolemy himself, uniformly adopted this second arrangement.

[73] *Almagest* IX, 1.

around which the epicycle's angular motion was uniform. Hindsight reveals this as a first glimpse of elliptical (and variable) planetary motion around an 'empty focus' – the hidden reef which wrecked all astronomical hypotheses before 1609.

Three points: (1) the *punctum aequans*, (2) the geometrical center of the deferent, and (3) the earth – these served as a center for angular velocities, a center for planetary distances, and the center for observations. After the Keplerian transformation these same computational points represented the two foci and the center of an ellipse, with *exactly the same ratios preserved.*

But such Ptolemaic violence to *the* principle of celestial mechanics – and allowing an epicyclic center to move regularly about a point *other* than its own deferential center *was* violent – seemed to Copernicus the scandal of Ptolemaic astronomy. His own theory removed it, but only at the expense of some observations.

Still, hindsight has been remarked as a deadly shoal for the historian and philosopher. Despite the formal connections between equant and ellipse, it is instructive to see the *punctum aequans* as Ptolemy's later critics saw it – a physically unintelligible notion which, even in a largely mathematical treatise, must constitute a flaw.

Ptolemy developed Apollonios' planetary theories considerably.[74] He detected that the angle formed (at the earth) by an epicycle's radius and the apsidal line was greater at apogee and smaller at perigee than the eccentric motion alone could account for. So, the conclusion had to be that the center of distances *was nearer to the earth than was the center* of uniform motion. Therefore, Ptolemy again locates on the apsidal line the infamous *punctum aequans* when dealing with the third lunar anomaly.[75]

So when (from earth) the planet is observed at *A*, *B* and *C* (in Figure 45), the time intervals between *A* and *B*, and between *B* and *C*, are equal; but it is clear that the planet's average velocity (i.e., the deferential motion of its epicycle) between *B* and *C* is almost twice what it was between *A* and *B*. As Ptolemy puts it himself:

Then we say that the whole plane revolves eastward in the direction of the signs [of the zodiac] about the center ... moving the apogees and the perigees one degree in a hundred

[74] Cf. *Almagest* III, 3; IX, 2ff.
[75] Cf. Figure 45, above.

years; thus the epicycle's diameter is in turn revolved regularly ... eastward in the direction of the signs at the rate of the stars' longitudinal return; and that at the same time it revolves the points on the epicycle, whose center is always borne on an excentric, and the star itself. And the star in turn moves on the epicycle, regularly with respect to the diameter always pointing toward [P], and makes its returns at the rate of the mean cycle of the anomaly with respect to the sun, moving eastward in the order of the signs at the apogee [of the epicycle].[76]

As we saw, when discussing the moon (and its perplexing double anomaly)[77], Ptolemy introduces this equant-idea which had such consequences for future planetary theory. Thus the first great fact of the heavens, which in the *Almagest* (II, 3) is enunciated as *the* principle of celestial mechanics – *that the stars all move uniformly on perfect circles around the centers of those circles* (the epicycle-centers moving regularly about the deferential-centers) – this is transformed by Ptolemy himself into a wholly different principle – *that a star* (or epicycle-center) *must move on a circle, and uniformly so, but about SOME other point which need not be the center of the circle it moves on*. Thus Ptolemy has the center of the moon's epicycle move regularly about the *punctum aequans*, and not about the center of its deferent.

The *punctum aequans*, and the interval between it and the deferent's center (the *equant*), were felt by most astronomers of the Late Medieval and Renaissance period to be somehow 'unphysical'. This mattered to some of them more than to others – and in different ways. Most could say: "Unphysical? So what? Astronomy is unphysical; all 'phenomenological' science is". A few demurred: "No discipline which is wholly phenomenological, wholly unconcerned with the reality 'behind' appearances, can be a full science at all."

---

[76] *Almagest*, IX, 6. This provides some answer to Geminus' question:

"Why, although the four parts of the zodiac's circle are equal, (does) the sun, traveling at uniform speed all the time, yet traverse the arcs in unequal times? For the hypothesis underlying the whole of astronomy, is that the sun, the moon, and the five planets move at uniform speeds in circles, and in a sense contrary to that of the motion of the universe ... no one would credit such irregularity (as we observed with the planets) even in the case of a steady and orderly man on a journey ... but when the stars, with their indestructible constitution, are in question, no reason can be assigned for swifter and slower motions."

(*Elements of Astronomy*, I.)

[77] *Almagest*, V, 2.

## Three Dimensional Variations of Ptolemy's Technique

Illustrations of his computational devices
Ptolemy's lunar theory

## Advantages of the Ptolemaic Theory

Its infinite flexibility (e.g. for distantial determination)
Allowance for variations in brightness

## Its Disadvantages

Its *ad hoc*, non-systematic character
The notorious *punctum aequans*

Most of Ptolemy's calculations are set in the plane of the ecliptic. But clearly, only motion in longitude can be determined in this manner. A looping reversal, such as those in Figures 8, 9, 10 and 11, requires some account of motion in latitude, as had been obvious to Eudoxos and Kalippus. Accordingly, Ptolemy tilted the orbit of Mars 1° from the ecliptic; Jupiter's was tipped 1°30', and Saturn's 2°30'. The apogees of these orbits were north of the ecliptic, but the epicycles themselves were tilted so that their planes were parallel to the ecliptic (see Figure 46).

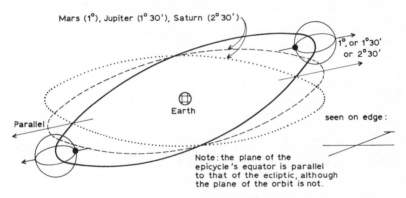

Fig. 46.   The three-dimensional 'tilt' of Ptolemy's epicycles.

Only thus could Ptolemy embrace the fact that these planets were observed farthest north or south of the ecliptic when at perigee in their epicycles.[78] This much would have seemed inexplicable to Ptolemy's predecessors and to some of his successors; they would have thought the most 'natural' arrangement to consist in the plane of the epicycle remaining parallel to that of the deferent.

[78] As we know, the epicycle of an outer planet is no more than the earth's annual circumsolar orbit transferred to the planet itself; it was thus quite correct of Ptolemy to make the epicycle parallel to the ecliptic.

Mercury and Venus presented special problems. Their orbits were only slightly inclined to the ecliptic. The epicycles, however, were *obliquely* inclined to the plane of the deferent[79] (see Figure 47).

Such devices permitted rough-and-ready geometrical representation for almost all naked-eye observations of Ptolemy's era. No wonder Ptolemy found latitudes so troublesome. In reality the lines of the planetary nodes pass through the sun; Ptolemy assumed that they passed

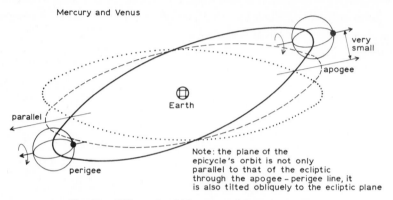

Fig. 47.   Oblique tilt of Mercury and Venus' epicycles.

through the Earth. Since the inner planets are enclosed by the Earth's orbit their motions in latitude appear more intricate. In this part of planetary theory the basic geocentric-geostatic supposition of Ptolemaic astronomy caused profound difficulty. To account for planetary latitudes remained the chief problem for all astronomers until Kepler.

Once all the appearances are saved by the consequences of the hypothesis ['any hypothesis possible'], why should it seem strange that such complications can come about in the movements of heavenly things?[80]

And in the introduction to his *Hypotheses*[81] Ptolemy even says:

I do not profess to be able thus to account for all the motions at the same time; but I shall show that *each by itself* is well explained by its proper hypothesis.[82]

[79]  *Almagest*, XIII, 2.

[80]  *Almagest*, XIII, 2.

[81]  Page 41, Halma.

[82]  Compare the following 'mere calculation':

"Next we take the same number ... to the table of anomaly. And, if the num-

Proclus[83] remarks that the epicycles and eccentrics are only a geometrically simple way of approximating actual celestial motions.[84]

The discrepancies between calculation and observation concerning the distance of the moon, e.g., showed the theory to be an ingenious formal aid to calculation, but certainly not a blueprint of the cosmos.[85]

Had he possessed an adequate algebra Ptolemy would surely have developed his theory in that less misleading medium.[86] Indeed, had Ptolemy written his *Syntaxis Mathematica* in Algebraic-Cartesian terms, no one could have imagined the *Almagest* to have been an attempt to

---

ber falls in the first column (that is, if it is not greater than 180°), then we subtract the corresponding degrees in the third column from the position of the mean course; but, if it falls in the second column (that is, if it is greater than 180°), then we add it to the mean course ..." (*Almagest*, III, 8. Compare also VI, 13, for a similar 'bookkeeper's-recipe' determination of the directions of rising and setting of the moon.)

Again:

"Multiplying the daily mean movements of each by 30 ...," "... multiplying also each of the yearly movements by 18 ...", "... now we shall again arrange, in the order of the stars for the ready use of each, the tables of these mean movements, in 45 rows and in lots of three like the others."

(*Almagest*, IX, 3.)

In fact, all of Section IV, Book IX of the *Almagest* gives, in its well-designed tables, an example of just what Ptolemy means to provide in his work. "... We may readily calculate from them the apparent passages for any time." (IX, 10.) And compare the Table, XIII, 8; also Table 10, XIII.

[83] *Hypotheses*, p. 151, Halma.

[84] Compare Ptolemy again:

"... being content for the present with the approximations which can be gotten each time from the catalogue or from the arrangement of the sphere ...". (*Almagest*, VIII, 6.)

"And we think it entirely proper to explain the appearances by the simplest hypothesis possible, so long as nothing perceptible appears contrary to this deduction." (*Almagest*, III, 1.)

This is a classic statement!

[85] "Let the concentric circle be conceived on the lunar sphere, and lying in the same plane with the ecliptic. And let another circle be conceived inclined to this one proportionately to the quantity of the moon's latitudinal course, and borne from east to west around the ecliptic's center at a regular speed equal to the excess of the latitudinal movements over the longitudinal ...". (*Almagest*, IV, 6.)

[86] We may anticipate the difficulties encountered by Newton's contemporaries; his theory, although expressed initially in cumbersome and opaque geometry, has a transparent algebraic equivalent. So also the work of Claudius Ptolemy.

*explain* anything – anymore than we now think an abacus, a slide rule, a computer or a system of formal logic, capable *per se* of explaining the physical world. Had Ptolemy's intentions been *that* clearly understood, however, he might never have had the effect upon Western astronomy that he did. The unprecedented success of his predicting devices, when coupled with his independent, but clearly articulated, faith in circular motion and in an immovable earth – all this led future philosophers to look upon the *Almagest* as a virtual source book of geocentric cosmology. The work can be read independently of all the Aristotelian frills which embroider the formal demonstrations.[87]

But, although Ptolemaic astronomy and Aristotelian cosmology, need never have been identified, as a matter of history they were. This strengthened each of them considerably, and delayed Copernicus' birth for 1300 years. Aristarchus was thus no Copernicus of antiquity. He was born too soon.*

---

[87] I make this claim despite the parallels between *Almagest*, (I, 7) and *Metaphysica* and *De Caelo* (II, 13–14) – and in other passages throughout the purely *cosmological* sections of these celebrated works.

* At this point in the MS, Professor Hanson's final corrections and revisions break off: the work had not been finally completed at the time of his death. The remainder of the text is as he left it, but there is no doubt that it is in most respects as he would have wished to see it published. Still, no doubt there are other small changes he would have wished to make to take account of recent scholarship.

# BOOK TWO

## PART I

## The Medieval Rediscovery of Ptolemy's Tool Box

*'The Ptolemaic System'*

It is not a *system* at all

*The Decline of Learning*

*The Transference of Greek Knowledge*

*The Early Medieval Cosmologies*

Historians of science invariably refer us to 'Ptolemy's System'[1]. Indeed, Toulmin and Goodfield, in their *The Fabric of the Heavens* set out Peter Apian's famous illustration of Aristotle's cosmology under the title 'The Ptolemaic World System'.[2] This is the same confusing conflation of Aristotle's cosmology and Ptolemy's astronomy that we have noted several times already. And Textbooks, like that of Baker[3] set out *drawings* of the 'Ptolemaic System' complete with the earth in the center and the seven heavenly bodies epicyclically arranged on their several deferents.[4]

Compare Margenau:

A large number of unrelated epicycles was needed to explain the observations, but otherwise the [Ptolemaic] system served well and with quantitative precision. Copernicus, by placing the sun at the center of the planetary universe, was able to reduce the number of epicycles from eighty-three to seventeen. Historical records indicate that Copernicus was unaware of the fundamental aspects of his so-called 'revolution', unaware perhaps of its historical importance, he rested content with having produced a *simpler* scheme for prediction. As an illustration of the principle of simplicity the heliocentric discovery has a peculiar appeal because it allows simplicity to be arithmetized; it involves a reduction in the number of epicycles from eighty-three to seventeen.[5]

---

[1] Cf. e.g., Dreyer, *op. cit.* Chapter IX, "The Ptolemaic System".

[2] Petri Apiani *Cosmographia*, per Gemma Phrysius Restituta, Antwerp, 1539.

[3] *Astronomy*, 5th ed., New York, 1950.

[4] Cf. also E. Rogers, *Physics for the Inquiring Mind*, Princeton 1960, p. 250; he commits the same error. The confusion is underscored by Holton and Roller (*Foundations of Modern Physical Science*, pp. 112ff) when they refer to 'Ptolemy's geocentric system' (§ 6.5), to the 'Success of The Ptolemaic System' (§ 6.6) and when they set out a diagram (Figure 6.10) purporting to schematize the Ptolemaic system. This latter surpasses even Baker's creation as a fictional invention: it corresponds to nothing ever set out in Ptolemy's *Almagest* as a 'system'. Indeed, there is very little anywhere in the entire history of science which is even closely depicted by this misleading diagram; remarkable, inasmuch as Holton and Roller are distinguished historians of science. So, the error they make cannot be charged to inexpertness. It is a genuine conceptual tangle, symptomatic of a misconception as generally widespread as are stories about the Galileo's Leaning Tower and the Falling Apples in Newton's Lincolnshire garden.

[5] *The Nature of Physical Reality*, New York 1950, p. 97.

In Section II of this Book, the misleading character of this way of talking, writing and thinking, will be set out in detail. But it is important here, for understanding the decline and transfer of later Greek learning, that the totally non-systematic character of Ptolemy's compilation of astronomical computing schemes be pointed out and stressed. Otherwise, it will be difficult to comprehend what precisely was being transferred back to the West as 'cosmology' and what as 'astronomy'. We have been at pains to underline the contrast between Eudoxos' collection of calculating techniques, and the systematic cosmology which Aristotle attempted to construct on their basis. The latter was seen to be largely inefficient as an aid to prediction, to navigation and to calendrical determinations. These objectives, however, dominated the undertakings of Hipparchus, Apollonios, and ultimately of Ptolemy himself. The 'big three' denied that any systematic understanding, or comprehensive explanations, would ever be the fruit of their industrious cerebrations. Whenever they lapsed into cosmological or metaphysical prose they simply parrotted the capital-lettered sayings of The Philosopher. But in their own technical astronomies there was no possible way of generating information about any one planet from nothing more than an exacting kinematical study of another. Knowledge of Mars' next stationary point could not be gathered from the *Almagest*'s analyses of the properties of the earth, Venus, the sun – or any other collection of bodies.

By confounding the cosmological and astronomical legacies of Greece, writers like Dreyer, Toulmin, Baker, Rogers and Margenau have made it difficult to appreciate that Copernicus' colossal contribution consisted in the *creation* of systematic astronomy. He was the inventor of a generalized astronomical computer, all of whose component parts were dovetailed together as parts of a whole. Too often it is made to sound as if Copernicus' only contribution was the 'shifting of the point of reference'.[6] He certainly did that! But he did this because no systematic astronomy was possible on the basis of Ptolemy's geostatic, geocentric coordinates. As we shall see, there never was any real competition between 'a Ptolemaic system' and 'a Copernican system'. The contrast was, rather, between an integrated, astronomical system, as against no system at all – a mere collection of computational tools. This point is

---

[6] This is most of Derek Price's thesis in 'Contra-Copernicus'.

central for the history of science: its significance has been obscured, however, by the fact that there *was* a quasi-philosophic competition between a geostatic-geocentric cosmology, and a heliostatic-heliocentric cosmology. This intellectual competition controlled all cosmological considerations for almost two thousand years. The contrast between systematic and non-systematic *astronomy*, however, affected Western thought for but a very short period of time, beginning only in the 16th century. Inasmuch as this distinction is itself informed by that between Explanation and Prediction, it will be important to bear it in mind as we proceed with the Medieval rediscovery of Greek studies of the heavens.

The thinkers within the early Roman Empire made very few contributions indeed, either to astronomical science or to cosmology. An influential compilation of all the discoveries, the arts and the sciences of mankind were set out by Pliny (23–79 A.D.) as a kind of textbook entitled *Natural History*. Indeed, this work became a standard repository for all later cosmological and astronomical writers. One of these compilers was Isidor of Seville (560–636 A.D.). In his *Etymologies* Isidor characterizes the universe as finite in size, as being but a few thousand years old and likely soon to perish. He envisaged the earth as shaped like a wheel encircled by the oceans. From thence outward to the concentric spheres bearing the planets and the stars, beyond the last of which was the highest heaven – the abode of all the blessed.[7]

Since it is no part of our objective to rewrite the Decline and Fall of the Roman Empire, it may be briefly noted that the metabolism of Greco-Roman learning, and of encyclopedia compilation, was deeply affected in the year 375 A.D., as was the entire Empire. It was then that the Huns poured into Europe through the natural gateway between the Caspian and the Urals. They drove the Goths, the Vandals and Franks headlong before them through the outlying provinces of the Roman Purple. In 476 A.D. the Purple Mantle was itself assumed by a barbarian

---

[7] It is significant that at about this same time Boethius, in *de Consolatione Philosophiae* (II, vii) wrote:

> "... the whole earth compared with the universe is not greater than a point, that is, compared with the sphere of the heavens it may be thought of as having no size at all."

This is the argument Archimedes tells of encountering in the works of Aristarchus, as we saw. But there is little likelihood that Boethius knew anything of either of those great Hellenistic thinkers.

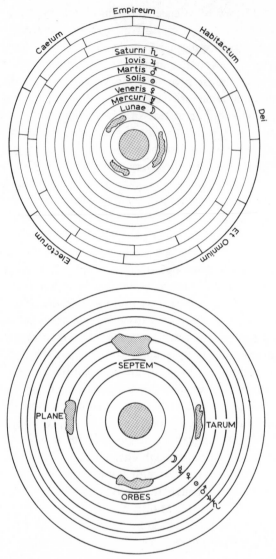

Fig. 48.   Late medieval representations of Aristotle's 'Cosmology' according to Apian.

chief. Ruin reigned. Within the new order there was considerable hostility toward any learning derived from the Greco-Roman world. Only commentators like Macrobius and Simplicius, as well as the encyclopedist

Martianus Capella, were left quietly to shelter and preserve the finest flowers of the Mediterranean's intellectual gardens; the barbaric, trampling 'anti-eggheads' stamped in from the North and from the East.

That so considerable a store of ancient learning was preserved was due to the rise of the monasteries and their associated schools; the first of these in Western Europe was Monte Cassino, raised by St. Benedict in 529 A.D. But amongst the Christians themselves there was as much enmity and savage hostility to astronomical science as ever there was within the paganry. The writings of Clemens Romanus (96 A.D.), Clement of Alexandria (200 A.D.), and Origen (254 A.D.) were more pictorially allegorical than technically astronomical. The firmament was viewed as a tabernacle, and all its constituent components were viewed through mythological lenses at least as chromatic as those which had perceived a detailed animated zodiac in the heavens. Within no time at all the sophisticated intellectual virtues of the Greco-Roman heathen came to be characterized as 'splendid vices'. Thus Father Lactantius wrote *On the Heretical Doctrine of the Globular Form of the Earth* (302 A.D.) in which he characterized it as a consummate absurdity that anyone should imagine crops 'growing downward' at the antipodes.[8]

In 360 A.D. Basil (ed. Gainier, Paris, 1721, tome I) wrote on the six days of creation. Much more sensible and controlled than Lactantius, Basil nowhere denies the results of Greco-Roman scientific inquiry; nonetheless he nowhere advocates them openly either. At one stage he endeavors to show that a Believer might even accept these scientific results without deep danger to his faith. The words of Genesis about the 'upper waters' confuse him, however, as they did the other Patristic writers: they were placed *above* the firmament to keep it cool and to prevent our world from being consumed by the celestial fire. A contemporary of Basil, Cyril of Jerusalem, stresses the necessary reality of these super-celestial waters.[9] Severianus, Bishop of Gabala, even more strongly stresses the literal reading of the first chapter of Genesis. Diodorus (394 A.D.), Theodore (428 A.D.), Philoponus and especially St. Jerome

---

[8] Copernicus much later remarks:
> "... it is not unknown that Lactantius, otherwise an illustrious writer ... speaks quite childishly about the earth's form ...".
> (*De Revolutionibus* ... Thorn, 1873, 7.21–23.)

[9] *Opera*, Oxford 1703, p. 116.

all wrote with strong feeling against the 'stupid wisdom of the philosophers'. For them Jerusalem was the navel of the earth.[10] Ambrose
of Milan (397 A.D.), and his most distinguished disciple Augustine
(430 A.D.) both speak more moderately of Greco-Roman astronomical
science. The latter, indeed, yields to it whenever Scripture does not un-
ambiguously pull him in the other direction. Indeed, St. Augustine was
a principle channel through which Mediterranean thought passed into
Latin Christianity; he had been profoundly influenced by Plato and by
Neoplatonists like Plotinus (270 A.D.). The *Timaeus* shaped much of
Augustine's cosmological outlook, and through him that of such thinkers
as John Scotus Erigena (877 A.D.).

Benighted writings such as those of Kosmas Indicopleustas will not
detain us. In our own day of scientific and astronomical sophistication
we need not hunt so very far for comparable violence and enthusiasm,
founded on ignorance and generated in the name of religion. Within the
civilized world now there is probably as much of this righteous know-
nothingism as there was when Kosmas wrote in 535 A.D. Fortunately
for us, we have in the 20th century a bit more of what might be called
'disinterested scientific inquiry' than was possible in the 6th century.
Johannes Philoponus, Isidorus Hispalensis and the Venerable Bede
sought to discuss more objectively than their immediate predecessors
the two great facts of the heavens. In the Latin treatise *De Natura
Rerum*[11] the influence of Pliny is most obvious. The spherical form
of the earth, the order of the seven planets circling 'round it, the sun being
described as larger than the earth, and similar facts – these are copied
verbatim from the ancient Roman's writings. The waters of Genesis,
however, are still superimposed on the stars. Boniface, in 748 A.D.
reported Virgilius of Salzburg to the Pope (Zacharias) for having here-
tically taught that there were "other people under the earth." The Pope
recommended expulsion of Virgil from the Church, had he really taught
such a vile thing.

Although under Charlemagne there was a revival of astronomical
learning in the 9th century, scholars such as Erigena were much out-
numbered by those who understood Christian innocence to be equivalent to Greco-Roman ignorance – by those who saw some divine virtue

[10] *Opera*, Paris 1704, II, p. 702 and 726.
[11] *Opera*, London 1848, Vol. VI, pp. 100–122.

in the preservation of their crania as *tabula rasa*. But by the time the mathematician Gerbert assumed the papacy as Sylvester the II (999 A.D.), the 'know-nothings' were losing their soporific effect on the re-awakening Western mind. Sylvester mentions Plato's *Timaeus*; he refers to Chalcidius, Eratosthenes and to other Greek teachers of a spherical earth.[12] Bede, and Erigena had already advocated the terrestrial rotundity. Adam of Bremen (1076 A.D.) indicates that he understands the causes of the inequality of day and night in different latitudes, as well as many other phenomena resulting from the earth's sphericity.

During the reign of Charlemagne the writings of Pliny, Chalcidius, Macrobius, and Martianus Capella were often copied, fairly well circulated, and rather widely read. In two works of the period (ascribed to Bede) several rotational inequalities are explained. Venus and Mercury are referred to as sometimes above and sometimes below the sun. Of the supercelestial waters, the other book says "... we think it was said more allegorically than literally". This second work shows awareness of the ancient differences of opinion concerning the solar orbit. The moon is there recognized as being *illuminated by* the sun, hence their orbits must be proximate. The sun is said to be eight times larger than the earth, and not made of fire alone.[13]

In all this, as it must be clear, what the Christian thinkers are dredging up from the dark depths of Mediterranean learning, derives exclusively from the cosmological side of our dichotomy between 'philosophical cosmology', vs. 'computational astronomy'. With the exception of isolated figures like Gerbert, there was barely enough geometrical horse-power extant even to understand the *Syntaxis Mathematica* (had it been available) much less to be interested in its contents. As Etienne Gilson has said of this period

... to understand and explain anything consisted, for the thinker of this time, in showing that it was not what it appeared to be, but that it was the symbol or sign of a more profound reality – that it proclaimed or signified something else.

After the bloody chaos that dominated the fighting for the spoils of Rome, the few dregs of ancient learning recovered were evaluated initial-

---

[12] Cantor, *Gesch. d. Math.* Vol. I, p. 811.
[13] The two works are *De Mundi Coelestis Terrestrisque Constitutione Liber*, and Περὶ διδάξεων *Sive Elementorum Philosophiae Libri* IV – both in *Ven. Bedae Opera*, 1612 tome I, pp. 323–344; tome 2, pp. 206–230.

ly in terms of what they could contribute to man's theological-allegorical comprehension of Western man's fierce and unfriendly environment. In such a context, it is not difficult to comprehend why the ancient wine-soaked cosmologies were in the ascendent over the icy clarities of Hellenistic astronomy.

The consequences of Erigena's conceptions were considerable. The early Master's devotion to the *Timaeus* rekindled the scientific interests of the successors of Charlemagne. The entire school of Chartres treated the *Timaeus* as a basic text. Gilbert de la Porree (1076 A.D.), Thierry of Chartres (1155 A.D.) and Bernard Sylvester (1150 A.D.) gave more attention to the Bible's scientific content than had been done since St. Augustine. Thierry, indeed, sought a rational explanation of the creation, and argued that *Genesis* could not be understood without a thorough gleaning in the quadrivium. In his account the Four Elements, so dear to the *Timaeus*, were built into a comprehensive cosmology of universal generation. It came to be argued that the motion of the heavenly bodies resulted solely from the universe, being spherical, having a proper motion of uniform eternal rotation in a circle around a fixed axle. The several spheres of the seven heavenly bodies revolved with different uniform velocities such as might represent the observed movements of those bodies; and in many other ways as well Plato's *Timaeus* affected the general philosophical outlook concerning the constitution of the universe. Roger Bacon himself, in the early thirteenth century, was lecturing on physics from the Platonic point of view. Being thus stimulated by such an 'objective' inquiry into the constitution of nature, it was no accident that it was at the school of Chartres that the Ptolemaic astronomy and Aristotelian physics were first welcomed into the northern renaissance.

It is well known that such treasures of Greco-Roman learning as had survived in the Rome of the Antonines – these picked up a value of their own as the Empire slid into decline. The manuscripts of the Mediterranean floated towards the Near East in the ships of soldiers and merchants. By the 800s, Venice, Naples, Bari, Amalfi, Pisa and Genoa were again initiating trade with the Arabs of the eastern Mediterranean. A knowledge of Arabic mathematics thus slowly began to distinguish the serious scholars at monasteries such as Monte Casino. Toledo fell to Alfonso XI in the eleventh century and became a Spanish center of translation, from Arabic into Latin. Uncompromising translation and transcription became a fundamental undertaking for learned scholars; consider the names – Adelard of Bath, Robert of Chester, Alfred of Sareshel, Gerard of Cremona, Plato of Tivoli, Burgundio of Pisa, James of Venice, Eugenio of Palermo, Michael of the Scot, Hermann of Carinthia, William of Moerbeke, John of Seville, Dominicus Gundissalinus – these are the most well known of a much larger squadron of scholars. Through their influence astronomy became enriched with the jewels of the Arabic language: 'Aldebaran', 'Altair', 'Deneb', 'Betelgeuse', 'Nadir', 'Zenith', 'Algebra', 'Algorism'. These words abound in the shiploads of Greek manuscripts which, in the twelfth century, ebbed back from the Near East to the northern Mediterranean. Plato's *Timaeus* was translated into Latin by Chalcidius in the fourth century; this explains why his influence was the earliest felt. Lucretius' *De Rerum Natura* was widely known in the twelfth century. Vetruvius' *De Architectura* was also well known at this time. Seneca's *Quaestiones Naturales*, Pliny's *Historia Naturalis*, Macrobius' *Insomnium Sciponis*, and Capella's *De Nuptiis Philologiae Et Mercurii* – these works were fairly well preserved in the monasteries during the darkest slope of the Post-Roman decline. Bede's *De Natura Rerum* constituted a high point in latter-day (original) Latin treatises. But later the astronomical tables of Al-Khwarismi (ninth century A.D.) were translated by Adelard (into Latin) – during the early twelfth century. The very important *Liber Astronomiae* of Alpetragius was translated from the Arabic by Michael the Scot (in Toledo) during 1217 A.D. This was the first really comprehensive rendering of the Aristotelian concentric system; as indicated earlier, this work had close connections with the later writings of Peter Apian, but *no* connection whatever with Ptolemy's *Syntaxis Mathematica*. Then Michael Scot translated Averroes' Commentaries on Aristotle's *De Caelo*; this again was an important contribution to the reviving medieval cosmology. In the twelfth century Aristotle's *Meteorologica* was translated from the Greek (by Henricus Aristippus) as was his *Physica*. In the thirteenth century almost all the works of Aristotle were rendered from the Greek into Latin, *via* new or revised translations by William of Moerbeke – a monumental achievement. And, from the point of view of this present work, the greatest achievement of all took place in the late twelfth century – when Ptolemy's *Almagest* was translated from the Greek (in Sicily), and from the Arabic (by Gerard of Cremona in Toledo). Moerbeke then translated Alexander of Aphrodisias' Commentary on Aristotle's *Meteorologica* from the Greek. The same translator rendered Simplicius' Commentaries on *De Caelo* and the *Physica*.[14]

The triumphs of Hellenic–Hellenistic cosmology and astronomy – products of the Golden Age of Greek thought – gilded also the Golden Age of the Antonines. In the tarnished and torn fragments of the empire, these nuggets were saved by scholars. The dedicated *literati* fervently wished to add luster to the Christian Faith that sprang from the volcanic ashes left by the lava-like lunges of the barbarians. The beauties of Greco-Roman studies of the heavens fell battered before the biceps of the embittered barbarian. The scraps were beatified by later Patristic thinkers as a kind of 'Christianized' cosmology. Men struggling for a comforting outlook during time of strife, went far in their imaginations to create a cosmos which blended at once the jewels of ancient learning with the overall structure and purpose of Christian doctrine and destiny. The cosmological speculations of the early Middle Ages, therefore, were uninformed with the minute analysis and computational accuracy which had been the glory of Greece from Eudoxos to Ptolemy. Just as today's floundering adolescent wallows in the 'big picture' and wonders 'what it is all about', not realizing until maturity that only accurate analyses of the details of a problem can comprise its real solution – so also early medieval re-discoverers of ancient learning re-cast the gems of Greece into a cosmological tiara of immense sweep and majesty. This they did long before the mathematical details of the work of Eudoxos, Kalippus, Hipparchus, Apollonios and Ptolemy could even be faintly comprehended. The result was an inspiring, uplifting and purple cosmical crown of super-explanation – *the* answer. Alas, the crown did not fit the heads of calendrical designers, cartographers or navigators whose immediate detailed problems required thinking caps of a different cut.

From all this we can see why it was Aristotle's cosmology which dominated the reviving European mind of the thirteenth century. The most advanced thinking of the period recognized that the cosmos was spherical and that it consisted in some large number of concentric spheres.

[14] Incidentally, the word 'algorism', and our own version – 'algorithm' – derives from the name of Al-Khwarizmi – the distinguished Arabic thinker of the ninth century.

The outermost was identical with the sphere of the fixed stars; the earth was fixed in the geometrical center of this entire sphere-cluster. As we saw, the sphere of the fixed stars was, by Aristotle, identified with the *primum movens*, the primary source of all movement within the universe.[15] Within this encompassing stellar sphere all the other cosmical spheres were nested and inter-nested – like the skins of a wild onion. But, unlike Peer Gynt's onion, Aristotle's cosmical onion had a heart, namely the Earth. And the Earth itself was spherically enveloped with a sphere of water, a sphere of air and a sphere of fire. Thence to the crystalline spheres delineated in Book One, Part I. Then the ultimate stellar sphere; and then nothing at all.

In such a cosmos every constituent had a natural place, a place 'fitting' to it. It also had a natural motion; if terrestrial, towards the center of the planet Earth, as is exhibited by the substance earth itself, and by water and by air. If a fiery substance, then the motion is still rectilinear, but now directed *away* from the Earth's center. If truly celestial, then the natural motion had to be perfectly circular and uniform, as with the rotation of a sphere on its axis. It was the sphere of the moon which divided the entire universe into everything celestial, or super-lunary – and everything terrestrial, or sub-lunary. Super-lunary bodies contain also a fifth element – a 'quintessence'. This was ingenerable, incorruptible and absolutely uniform; its natural motion was, as indicated, perfectly circular – the only kind of motion that could persist eternally in a finite world. Plato had characterized circular motion as the most perfect of all; his restriction of all heavenly bodies to this motion and this alone actually dominated all astronomy and cosmology until the late sixteenth century. All motion in the universe was communicated from the *primum mobile* to the inner-sphere by mechanical contact, the exact details of which were not, as we have seen, pellucidly clear. As Crombie says of the dominating contention of this period "... all change and motion in the universe was ultimately caused by the primum mobile."[16]

Now, with the twelfth century translations of the *Almagest* sailing straight into these cosmological currents, echoes of earlier controversy began to reverberate in all the learned circles of Europe. Those who called themselves 'physicists' spoke out as Aristotelian cosmologists.

[15] Compare Apian's version in Figure 48.
[16] *From Augustine to Galileo*, p. 55.

Those who called themselves 'mathematicians' advocated Ptolemy's astronomy. The prose-laden dynamics of The Philosopher was pitted against the mathematically-articulated kinematics of The Astronomer. The contest was, as we saw, initiated by Michael the Scot with his translation of Alpetragius' *Liber Astronomiae*; in that work one sees an attempt to bolster Aristotle's work as a *predictive astronomy* as against the much more accurate calculations of Claudius Ptolemy. At first Aristotle's ingenious reconstruction and generalization of the calculations of Eudoxos and Kallippus constituted the ultimate in medieval thinking. And, if one consults our earlier Figures (14–22) it can be appreciated that these early ideas were exacting enough.[17]

In the early thirteenth century William of Auvergne suggested separating the *primum mobile* and the sphere of the fixed stars. And beyond the *primum mobile* he set an absolutely immobile Empyrean – where the saints resided. So, whereas some earlier thinkers had conceived of seven major planetary spheres and an eighth – the *primum mobile* – William suggested a ninth and a tenth! Here we see the injection of Christian theological dogma into the rediscovered cosmologies of the pre-Christian era.

The major weakness of the Eudoxian-Kalippian-Aristotelian technique was the same in the thirteenth century A.D. as it had been in the fourth century B.C. The assumption that each heavenly body remained at a constant radial distance from the earth was abruptly jolted by the immense variations in the apparent brightness of certain planets (e.g., Venus, cf. Figure 26), variations in the apparent diameter of the moon, and the occasional occurrence of an annular solar eclipse. The techniques

---

[17] Incidentally, Crombie (*op. cit.*, p. 57, Figure 2) has certainly erred in his depiction of the Eudoxian-Kalippian-Aristotelian planetary technique. He identifies the outer sphere of a given planetary-cluster as the 'stellar sphere'; for Aristotle this outer sphere, e.g., ♂ α, only "has the motion of the fixed stars" – it is not identical with that sphere. Crombie's second sphere is tilted, one may suppose, 24° from the axis of the stellar sphere. But then another sphere (the third) inside that one is just slightly offset from the poles of the second sphere – whereas, for Eudoxos, Kalippus and Aristotle this third sphere is supposed to have its poles fixed in the *equator* of the second sphere. The fourth sphere should be set on an axle slightly askew from the axis of the third sphere – just enough to give the required oscillations in latitude of the planet through the zodiac. It is on the *equator* of this last sphere where the planet is meant to be fixed. Crombie's planet is *not* set in that equator; the planet-carrying sphere is *not* set at the zodiacal limits of the planet's course along the ecliptic; his third sphere is *not* fixed in the equator of the second one; and his first sphere should *not* be identified with the sphere of the fixed stars. All this is totally wrong. The correct configurations are set out above in Figures 14, 15, 16, 17, and 23.

of Hipparchus and Ptolemy were just slightly superior in this respect. It was no part of their schema that the planet had to remain at a constant distance from us; only their perfectly uniform and circular motion was required. It was these major facts of observation, incompatible with the consequences of Aristotle's 'astronomy', that slowly swung serious astronomical speculation toward the techniques set out in Ptolemy's *Almagest*. Again, Ptolemy's clear distinction between a universal, explanatory cosmological system and his own collection of geometrical devices, by appeal to which one might effect *apparentias salvare*, this distinction becomes vital in the prosecution of late medieval mensurational astronomy. For only by conceding that he was merely seeking to 'save appearances' (for the sake of navigation or calendry) could a medieval Schoolman justify giving his attention to such a 'non-Aristotelian' undertaking as is Ptolemy's astronomy. The theory of epicycles was incompatible with Aristotle's dictum that perfect circular motion required a fixed center around which to revolve. Ptolemy's explanation of precession, as we saw, requires that the stellar sphere have two different (and even opposed) motions at the same time, this being clearly in conflict with Aristotle's principle that contradictory attributes cannot inhere in one substance at one time. But, although deficient from the point of view of its cosmological commitments, Ptolemy's techniques were vastly superior as a precise mathematical description, and prediction of the observed phenomena. So Ptolemy's implicit distinction between 'mere astronomical prediction' and 'real cosmological explanation' assumed profound importance. Thus St. Thomas Aquinas in the *Summa Theologica* (I, q. 32; Art. 1) distinguishes hypotheses which are philosophically true from those which merely fit the facts.

> ... a system may be induced in a double fashion. One way is for proving some principle, as in natural science, where sufficient reason can be brought to show that the motions of the heavens are always of uniform velocity. In the other way, reasons may be adduced which do not sufficiently prove the principle, but which may show that the effects which follow agree with that principle – as in astronomy a system of eccentrics and epicycles is posited because this assumption enables the sensible phenomena of the celestial motions to be accounted for. But this is not a sufficient proof, because possibly another hypothesis might also be able to account for them.

In the later thirteenth century Bernard of Verdun and Giles of Rome argued that astronomical hypotheses should be constructed solely to explain the observed phenomena: experimental evidence should be the

only way of settling the controversy between the Aristotelian 'dynami-
cists' and Ptolemaic 'kinematicists'. The views of these two thinkers
are important: they are arguing, Hempel-style, that no orthodox philosophical cosmology (not even Aristotle's) can be acceptable as a satisfactory description of the facts. They are opting for one set of theoretical
criteria to adjudicate the scientific worthwhileness of *any* study of the
heavens. They seek to rid the field of knowledge of its double standard –
one for philosophical cosmologists, and another for computational
astronomers.[18]

Before the fourteenth century Aristotle's cosmology had been wholly
discarded in the Schools of Paris. The devices of the *Almagest* have been
altogether accepted in the light of observational experience.[19] Attempts
were made to mesh the two approaches, as when Ptolemy's epicycle and
eccentrics were construed as solid quintessential spheres. But such interpretations were less important than the calculational power afforded
by the *Syntaxis Mathematica*. Even St. Thomas[20] shows a clear understanding of the controversy. He is well-acquainted with the work of
Simplicius, of Plato, of Ptolemy and of all the later Patristic commentators. He notes the reasons for substituting epicycles and eccentrics for
the homocentric spheres postulated by Aristotle. He remarks the reason
for placing another sphere beyond that of the fixed stars, namely the fact
of the proper motion of the fixed stars themselves. And in arguing for
the geostatic geocentricity of the universe he sets out Ptolemy's detailed
arguments verbatim.[21]

Contemporary with the work of Thomas is *The Sphere*, written by
Johannes de Sacro Bosco (John of Holywood). He cursorily refers to
Ptolemy and Alfargani during his truncated descriptions of the equants,
deferents, epicycles, and eccentrics. Indeed, Sacro Bosco gives an elementary sketch of the Ptolemaic technique. True, his knowledge seems
to be rudimentary and very second hand. Usually he simply states the

---

[18] Only with Newton, in the seventeenth century, is their insight at all realized.

[19] Given this battle, and this signal victory, how doubly wrong it is of Toulmin to characterize Peter Apian's *Cosmographia* as 'the Ptolemaic world system'. Crombie is careful to
characterize Apian's diagram as a depiction of 'Aristotle's cosmology' – although Crombie
carelessly refers to 'the Ptolemaic system' in many pages of his *From Augustine to Galileo*.
[20] *Opera Omnis*, C. III, 'Commentaria In Libros Aristotelis De Coelo Et Mundo', Rome
1886.
[21] II, xxvi, p. 220b.

problems and their summary solutions, without any of the proofs, as these are set out in the *Almagest*. He copies even the mistakes of both Alfargani and Albattani, and in general, he seems to be setting out a crude popular account of Ptolemy's work, suggestive of the treatment General Relativity might receive today at a Junior College. But that Sacro Bosco's work was 'popular' is an indirect reflection of how the battle fared between the prose-drenched cosmologies and the geometrically-informed astronomies of the late thirteenth and early fourteenth centuries. Indeed a host of further writings on 'The Sphere' were undertaken simultaneously with, and subsequent to, Sacro Bosco's work. A lot of this was simply crude spherical geometry or erroneous trigonometry – all directed toward a study of the heavens on high. But it is significant that, by this stage, such studies are felt to be more effectively carried out in mathematical terms than in physical ones.

But although the non-systematic *Almagest* was definitely in the ascendant above the over-systematic cosmologies of Aristotle, stellar controversy continued. In the *Ludicator Astronomiae*, Peter of Abano actually suggests that the stars are not borne on a sphere at all, but move freely in all directions in outer space. There were some difficult-to-comprehend attitudes rather like this expressed in very ancient times (e.g., by Empedocles) but this one seems to be calculated as an explicit departure from orthodoxy. Indeed, by the end of the thirteenth century the Franciscan, François de Meyronnes, considered the logical consequences of letting the earth, rather than the heavenly spheres, revolve. Independently, this inquiry was later taken up by Nicole Oresme. Doubtless, the speculations and reflections of Cusa and Oresme constitute the attainment of a high plateau in the struggle for the freedom with which the Western mind was prepared to entertain and elaborate even hypotheses which at first seemed unpromisingly heretical.[22] It is quite possible that the radical new hypotheses of Cusa and Oresme, in the late fourteenth and early fifteenth centuries, were suggested originally by the ancient Greek speculations – notably the semi-heliocentric suggestion of the fourth century B.C. made by Herakleides of Pontus. In this so-called 'Egyptian system' Venus and Mercury revolved around the Sun

---

[22] The explorations of Dr. Edward Grant in this connection are anxiously awaited by all scholars.

while the Sun itself revolved around the Earth. Knowledge of this 'system' percolated into Western Europe through the writings of Macrobius and Martianus Capella. The connections between these ideas and the Tychonic system of the sixteenth century will be examined later.[23]

Another thirteenth century scholar, with powers comparable to those of Aquinas himself, was Roger Bacon (1294 A.D.). In his striking *Opus Majus* Bacon reveals himself as being deeply immersed in the Greco-Arabic literature, much more than most of his contemporaries; it is probable that Roger Bacon could actually follow the detailed mathematical arguments of Ptolemy. Ptolemaic ideas are religiously followed through most of Bacon's inquiries. Bacon had no trepidation about pointing out the immense perplexities in any literal interpretation of Scripture. He singled out the first chapter of Genesis, the references to Joshua, passages in Isaiah and remarks of St. Jerome as being clearly in conflict with the eighth book of the *Almagest* – a fact which he takes to reflect adversely on Scripture rather than on Ptolemy.

Learned planetary theorists, like Thomas and Bacon, were not alone in their quickening interests toward the Ptolemaic techniques. The Church was continually in need for some reliable tables devoted to the calculation of holy days and religious feasts – Easter in particular. Calendrical computation required some precise and accurate descriptions and predictions in order even to approximate to what was required by tax-collectors, soldiers, navigators, farmers and governors. Even the objectives of astrological prediction(!) required the increased accuracy afforded by the computing techniques compiled within Ptolemy's *Syntaxis Mathematica*. The cosmological implications built into these calculating devices were no more relevant to the immediate problems in the thirteenth and fourteenth centuries, than are the geocentric implications built into our own Nautical Almanac. Indeed, the Nautical Almanac is virtually a text book of geocentric astronomy; but this is a fact of which our seamen and fishermen are aware.

The eleventh century calendar of Khayyam was nearly as precise as anything produced until 1582; this revealed the superiority of Arab instruments, observations, tables and maps. Later in the eleventh century Walcher of Malvern traced an eclipse of the moon while in Italy; a friend

---

[23] The work, and ideas, of Aristarchus of Samos was completely unknown throughout the medieval period.

observed the same eclipse in England. By comparing the times of these observations Walcher was able to determine exactly the longitudinal difference between the two points. Several uses of the astrolabe served well in determining latitudes. Indeed, the astrolabe went through an intricate evolution of refinements and improvements during the thirteenth and fourteenth centuries. Geoffrey Chaucer himself wrote an excellent treatise on *The Astrolabe* in the late fourteenth century. Derek Price has even presented a powerful argument to the effect that Chaucer authored a detailed text on the construction of an equatorie – an observational instrument of surprising accuracy. Observational and practical astronomy was therefore becoming ever more important in the high medieval period.

Then, as now, there was a constant need for *numbers* in description and prediction. And where this could be achieved only by appeal to techniques which rested on 'uncomfortable' assumptions, it was universally moved that these assumptions were made 'for calculation merely'; the major objective remained *apparentias salvare*.

Thus it is that men within *literae humaniores* were drawn at once toward two slightly opposed goals. In Chaucer's *Equatorie of the Planetis* there is manifested the human desire to *control* nature – by describing its behavior with relentless accuracy, and by predicting what will occur next.

In Dante's *Divine Comedy*, on the other hand, we see again the human desire to *understand* our natural context; to explain the nature of the world into which we have been placed, from the Firmament on high to the Nether regions below.

These perpetual objectives developed in the medieval mind as if they were constitutionally dissimilar. In the thirteenth and fourteenth centuries we sail through the intellectual Antipodes diametrically opposite to Hempel's analysis of the explanation-prediction symmetry. Because, for Aquinas, Bacon, Chaucer, Cusa, Dante and Oresme – amongst many others – it appeared that in matters celestial one could have *either* mathematical description and prediction, or cosmological understanding and explanation – but never both at once. One could partake of The Philosopher's sweeping and comprehensive vision of the universe as a whole, *or* one could settle down to the more work-a-day uses of astrolabe, gnomon and backstaff, observations with which might then be harnessed

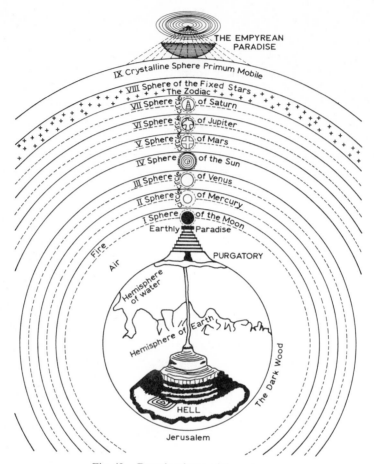

Fig. 49.   Dante's scheme of the universe.

to the computing techniques of the navigator's friend, the *Almagest*. The sharp distinction between these two types of inquiry, obvious as it was in Aristotle and Geminus is stressed again in the works of Thomas; it even constitutes a major dichotomy throughout the sixteenth century, as our discussion of Osiander's *Preface* to Copernicus' work will show.

But this contra-Hempelian situation was far from being a bad thing! It may be said to have prepared the way for Copernicus, and ultimately for Newton – within whose works Hempel's account is paradigmatically illustrated. Because, it was this very separation of explanation – understanding from description – prediction which made it possible for thinkers like Cusa and Oresme to develop their boldest speculations. For, just as the medieval distinction between Faith and Reason allowed Schoolmen to explore the logical structures of argument (even arguments for the existence of God) without 'endangering' their religious beliefs, so also the sharp separation between explanation and prediction allowed scholars of the fourteenth and early fifteenth centuries to nod in a fully affirmative way towards their theologically-impregnated cosmologies, while nonetheless exploring to the hilt both the accurate predictional machinery of Ptolemy, as well as the consequences of alternative astronomical hypotheses *qua* hypotheses. Thus Oresme:

> I suppose that locomotion can be perceived only when one body alters its position relative to another. Therefore, if a man in a smoothly riding boat, A, which is moved either slowly or rapidly, can see nothing but a second boat, B, which moves in just the same way as A, ... then I say that it will seem to him that neither boat is moved. And if A is at rest and B moves, it will seem to him that B moves; and if it is A that moves and B is at rest, it will still seem to him that A is at rest and that B is moved ... Therefore I say that if the higher [or celestial] of the two parts of the universe mentioned above were today moved with a diurnal motion, as it is, while the lower [or terrestrial] part remained at rest, and if tomorrow on the contrary the lower part were moved diurnally while the other part, i.e., the heavens were at rest, we would be unable to see any change, but everything would seem the same today and tomorrow. It would seem to us throughout that our location was at rest while the other part of the universe moved, just as it seems to a man in a moving boat that the trees outside the boat are in motion.[24]

Oresme, in this passage, is not arguing that the earth *does* in fact move. But, because of the sharp contrast between his theological-cosmological explanations (to which he is committed as a Believer) and the logical structure of alternative descriptions (which the license to explore the consequences of reason grants him) – Oresme can here examine what is entailed by the supposition, and the possibility, that the earth *could* move.

The upshot of inquiries such as those of Oresme, Cusa and Buridan was that it was seen clearly that, from a logical point of view, several *alternative* astronomical hypotheses were feasible for "saving the celestial appearances". It was one's religious commitments which singled out a preferred hypothesis because of its cosmological consequences. Nonetheless, the cleavage between cosmological explanation and astronomical prediction now permitted scholars to explore those astronomical alter-

---

[24] *Le Livre du Ciel et du Monde*, IV, p. 272.

natives in increasing detail, notwithstanding their suspect cosmological overtones. Without such explorations – without seeing the efficient ways in which these 'alternatives' succeeded in accounting for celestial phenomena – the brilliant, late-medieval proposals of Copernicus himself might never have been undertaken.

Bracketed with Oresme's free inquiry into astronomical alternatives are his trenchant criticisms of the more mundane features of Aristotle's *Physics*. As is well known, The Philosopher felt motion to be possible only when some 'force' was actively moving the object in question. The natural state for all terrestrial objects was one of rest. Only by violent dislocation could the coordinates of terrestrial objects be altered. The natural state of celestial objects, on the other hand, was to *remain* in perfectly circular motion, as necessitated by the dynamic influence over the heavens of the *primum movens*. Characteristically different types of motive law thus divided the universe, much as the lunar sphere divided the universe. But throughout the twelfth, thirteenth and fourteenth centuries, Aristotle's account of terrestrial motion was a continual source of conceptual perplexity. If a moving object moved only because is was at that time being actively pushed, how could one explain the lengthy flight of an arrow which, after the initial impetus imparted to it from the bowstring, seemed to soar through the heavens – without the continued operation on it of the bow, or the archer, or anything else. Aristotle had suggested that the air cleft atwain by the arrowhead somehow circled around behind the shaft and pushed it forward during every moment of its flight. But what then of a javelin, needle-sharp at *both* ends, which could be thrown half a stadium (half of 516.73 ft.) by any Greek athlete, or any medieval soldier? No; Aristotle's "continually-forced-if-moved-at-all theory" began to give way to the view that the spearman, or the archer, contributed something to the projectile once and for all at the beginning of its trajectory – something which was completely responsible for the ensuing flight. This idea was soon refined into the Impetus theory of motion, a distinguishing and characteristic conception of the later Parisian school of physicists.

Buridan, a member of that school, and Oresme's teacher, generalized this theory of motion not only to account more satisfactorily for the moving of terrestrial object, but also to deal with celestial motions. Thus he says:

... Since the Bible does not state that appropriate intelligences move the celestial bodies, it could be said that it does not appear necessary to posit intelligences of this kind [as had been done in Aristotle's cosmology]. For it could [equally well] be answered that God, when He created the world, moved each of the celestial orbs as He pleased, and in moving them He impressed in them *impetus* which moved them without His having to move them anymore except by the method of general influence whereby He concurs as a coagent in all things which take place. Thus on the seventh day He rested from all work which He had executed by committing to others the actions and passions in turn. And these impetuses which He impressed in the celestial bodies were not decreased nor corrupted afterwards, because there was no inclination of the celestial bodies for other movements. Nor was there resistance which would be corruptive or repressive of that impetus.[25]

In this quotation three monumentally important intellectual drives are at work. One; Buridan is entertaining an alternative to the 'accepted' cosmological commitment of Aristotle; this he appears to do for astronomically cogent reasons. Two; he is bracketing both celestial and terrestrial motions under a single set of laws, curtailing thereby the traditional *ad hoc* tendency to account for new phenomena in the most readily available terms, regardless of how all the resulting accounts might mesh together. Three; Buridan argues that the celestial bodies move in uniform perfect circles "because there was no inclination of the celestial bodies for other movements"; in short, the bodies move as they do because it is in their nature to preserve what motion has been imparted to them, and "they are free of later impressed forces".

Nicole Oresme adds intellectual impetus to these thrusts of imagination already unleashed by his teacher. The urge to bracket, dovetail and unify one's accounts of phenomena, and to link terrestrial physics and theoretical astronomy – these along with the Schoolmen's new-found freedom to explore hypothetical alternatives – directly paved the high road leading toward 'the Copernican System' – the capstone of medieval astronomy.

---

[25] From Marshall Claggett's *Selections in Medieval Mechanics*, p. 40, Buridan's *Quaestiones*.

# SUPPLEMENTARY MATERIAL FOR BOOK TWO, SECTION A

The earliest tracts of Aristotle to reach the reviving West were the logical works – the *Logica Vetus* – rendered into Latin by Boethius during the sixth century. A good deal of Boethius' work in mathematics and astronomy, as well as his *Commentaries* on the logic of Aristotle and Porphyry, were available in the earliest middle ages. In the early twelfth century Aristotle's *Posterior Analytics* – part of the *Logica Nova* – were translated into Latin. Then in 1126 Adelard of Bath translated the fifteen books of Euclid's *Elements* from the Arabic into Latin. In short, logical studies, and the analysis of cogent argumentation, were themselves responsible for much of the intellectual quickening which we now recognize as the medieval revival of learning. The Schoolmen were very early aware of the logical distinctions to be drawn between *knowledge of facts* and *appraisals of valid argument*. They knew the difference between factual proofs and logical consistency. And they realized that consistent consequences might be shown to follow from factually untrue premises. They could also detect logical inconsistencies in arguments generated invalidly from factually true premises. These distinctions, pellucidly clear in Aristotle's own works, reinforced the cleavage between (1) the dogmatic Truths which had been injected into the 'Christianized' cosmology of The Philosopher, and (2) the logical and mathematical consistency so obvious in the treatises of Ptolemaic astronomy. It often happened that scientific commentators like Adelard of Bath and Hugh of Victor, as well as Anselm, Richard of St. Victor and Abelard sought to set out their subject matters in accordance with a kind of mathematical-deductive-logical method of exposition. Thus, to anticipate Galileo, Salviati is made to say of the Ptolemaic theory that

It satisfied an *astronomer* merely *Arithmetical*, yet it did not afford satisfaction or content to the *Astronoma-Philosophica*.[26]

The revival of sharp and sure interest in mathematics was marked by

[26] *Two Principles Systems*, The Third Day.

the wide circulation of the *Geometry of Boethius*. The writings of Cassiodorus, Isodore of Seville, Gerbert, Adelard, Françon of Liège, Rainbaud of Cologne and Radulfus of Liège – although this was largely rudimentary and unoriginal in character – constituted a new metabolism for mathematics throughout the twelfth century. The great name in the early thirteenth was Leonardo Fibonacci of Pisa: his knowledge was largely based on Arabic sources, as well as on Euclid, Archimedes, Hero and Diophantus, the great Greek algebraist. Fibonacci had a clear interpretation and understanding of generalized proof and of negative magnitudes, he algebrized many geometrical problems and actually was familiar with quartic equations. Jordanus Nemorarius developed further the tradition of Nicomachus and Boethius in number theory. He was quite familiar with linear and quadratic equations and knew a fair amount about the properties of stereographic projection. Campanus of Novara was a student of 'continuous quantities'; he also was thoroughly familiar with irrational magnitudes. The names of John Maudich, Richard of Wallingford and Levi ben Gerson as well as Thomas Bradwardine, Albert of Saxony and ultimately Nicole Oresme himself, would be prominent in any history of serious mathematics. All of which is calculated to show that the idea of a really demonstrative science was again quite viable by the thirteenth century. A heightened interest in *demonstration* – in 'deductive unpacking' – had a very special interest for those Scholastic minds who were not content simply to rehearse *seriatim* the unnumbered and unconnected details of Church dogma. Since these latter were always granted to constitute the *ultimate Proof*, monastic scholars could feel free (in virtue of the cleavage between truth and consistency – factual knowledge and valid demonstration – logic, astronomy and cosmology) to exercise their talents in logic, argumentation, demonstration, calculation and computation.

Robert Grosseteste, as has been hinted by Harrison Thompson and established by Alistair Crombie, erected his entire philosophical outlook on Aristotle's distinction between one's knowledge of a fact, and knowledge of one's reasons for asserting that fact. His special interests in argumentation took him to the heart of verification, of deduction, and of the problems of scientific method. Grosseteste's enthusiasm for analyses of the structure of scientific argumentation led to the freer kinds of inquiry and marked departures from Aristotle that one begins to find in John Duns Scotus (1308 A.D.) and William of Ockham (1349 A.D.). Scotus drew a clear distinction between causal laws and empirical generalizations – a distinction not unrelated to the contrast between 'truths of reason' and 'truths of fact' (to use Hume's terminology) so central to the understanding of Aristotle's logical work (and all medieval commentaries thereon). And empirical generalizations began to look more like purely accidental correlations, factual summaries, to which there were many possible theoretical alternatives. The causal law, however, came to be seen as delineating a kind of structural connection between events, a connection which seemed to follow from understanding *the nature* of the phenomena in question. On the basis of a merely empirical generalization one could describe and predict, but never explain; just as we might be prepared for a Kansas wheat failure on the basis of an unprecedented storm of sun spots – 'prepared' but not 'convinced'. Or, we might predict that the next white, blue-eyed tomcat we encounter will be deaf. Our reasons in both cases are that we have generalized such a correlation from a large number of similar joint-occurrences observed in the past. But we do not in these cases pretend to *understand* the correlation, as would be the case with a genuine causal connection, e.g., a cueball nudging the eightball into the side pocket. Scotus made much of this distinction between generalizations and laws, and he aligned it with our already-familiar distinctions between predictions and explanations, mathematical description and physical understanding, astronomical computation and cosmological insight. One of Scotus' objectives was to examine in greater detail the theory of induction set out by Aristotle in Book Two, Ch. 19 of the *Posterior Analytics*, wherein the avowed intention is to examine the conceptual anatomy of laws of nature. (Would that subsequent philosophers of science had been as steadfast in their examinations and their distinctions.) The logical and methodological structure of a detailed theory of induction is thus generated in these thinkers, in the course of setting out which Aristotle's own views are regularly challenged and often shattered – by inflexibly consistent applications of the very logical tools The Philosopher himself honed for posterity. On a wide range of logical and methodological issues William of Ockham 'put the question' to Aristotle. Again, in the *Summa Totius Logicae*, III, Section 2, Ch. 10, Ockham takes the distinction between empirical regularities and genuine lawlike connections and builds it into the foundations of a theory of induction which invites contrast with Aristotle's own. Ockham's conceptual separation of 'rational science' and 'empirical science' ends in some passages reminiscent of the writings of Whewell, Mill, Pearson, Russell and Reichenbach. Thus:

> ... something [is] the immediate cause ... when it is present [and] the effect follows – and when not present, all other conditions being the same, the effect does not follow.[27]

The cosmological-astronomical implications of this heightened criticism of Aristotle were considerable. In a sense, The Philosopher's cosmology could *not be said* to provide any real understanding of the universe when contrasted with the 'merely astronomical' alternatives. Because the former could not itself be shown founded on any sound theory of inductive knowledge. Even the ancient dictum 'a plurality must not be asserted without necessity'[28] would have an obvious negative-effect on the prose-laden, theologically embued, decorative Aristotelian cosmologies. The very superfluity of terms in some of these made them vulnerable to the insight of Nicolaus of Autrecourt to the effect that:

> From the fact that one thing is known to exist, it cannot be evidently inferred that another thing exists ... whatever conditions we take which may be the cause of any effect, we do not evidently know that, when conditions are posited, the effect posited will follow.

Aristotle's dynamics were systematically challenged and attacked as well, in such remarks as that "... This moving thing and the motion cannot be distinguished",[29] and

> $X$ is moved means $X$ is at $A$ at time $t$, $X$ is separated from point $B$ at time $t$, $X$ is at $B$ at time $t_1$, and separated from $A$.[30]

Thence to Buridan:

> "In moving a body, a mover impresses on it a certain *impetus*, a power capable of moving it in the direction the mover determines – upwards, downwards, sideways or in a circle ... It is by this *impetus* that the stone is moved after the thrower has released it; it is the resistance of the air and of gravity, which inclines the stone to move against the direction towards which the impetus propels it – this impetus is thus continually weakened ... At length it is so diminished that gravity prevails and moves the stone towards its natural place ... Anyone who wants to jump far, draws back a long way so that he can run faster

[27] *Super Libros Quatuor Sententiarum*, I, dis 45, question 1d.
[28] *Quod Libeta Septem*, q5, question 5.
[29] Ockham, *Super Quatuor Libros Sententiarum*, II, question 26m.
[30] *Nicolaus of Autrecourt*, cf. J. R. Weinberg, p. 168

and acquire an *impetus* which, during the jump, carries him a long distance. Moreover, while he runs and jumps he does not feel the air behind him moving him, but he feels the air in front resist with force. Also, since the Bible does not state that appropriate [angelic] intelligences move the celestial bodies..." etc. cf. *Quaestiones Octavi Libri Physicorum*, Book 8, question 12; and in *Quaestiones in Libros Metaphysicae*, Book 12, question 9, Buridan argues "the *impetus* would last indefinitely were it not diminished by a resisting contrary force, or by an inclination to a contrary motion; and in celestial motion there is no resisting contrary ..."

Here again we find terrestrial and celestial mechanics meshed into a single system. In all this Buridan was followed by Albert of Saxony and Nicole Oresme; indeed, the theory of impetus virtually 'took over' physical thinking in the fourteenth, fifteenth and sixteenth centuries in northern Europe.

Buridan's impetus theory deeply affects Oresme's *Commentary on Aristotle's De Caelo*. Oresme entertains the hypothesis of the earth's rotation, and fixes that body centrically within an otherwise motionless universe. He then counters the standard Aristotelian and Ptolemaic 'physical' objections to the earth's rotation, and he does so in ways which anticipate the later arguments of Copernicus and Bruno. He says:

... all the appearances are saved by a small operation – the earth's daily rotation ...; ... and considering everything that has been said, one can then conclude that the earth is moved and the sky not, and there is no evidence to the contrary.[31]

Thus for all these thinkers – Buridan, Oresme and Albert – cosmo-logical-astronomical writings were almost exclusively exercises in logical derivation. Given the distinction between truth and consistency – between facts and validity – they are intellectually enthusiastic about tracing unflinchingly the consequences of hypotheses which (on other, theological, grounds) they would be most eager to deny. But the histori-cally significant point here is that alternative hypotheses *are* thoroughly explored, semantically unpacked and logically decomposed – for the first time in ages. It is within this highly critical climate, in which hypo-theses alternative to the great Aristotelian tradition are developed in order to probe the weaknesses of The Philosopher's cosmological pronouncements, that the wedding between cosmology and astronomy is at last brought about by Copernicus – a thinker very much in the immediate tradition of Buridan, Albert, Oresme and Nicolaus of Cusa.

[31] Book II, Ch. 25.

Cusa (1464 A.D.) actually proposed that the sphere of the fixed stars twisted twice about *its* axis for every single rotation of the earth around its own. But this idea, as with the often-formless suggestions of Nicholas' predecessors, offered no genuine mathematical alternative to Ptolemy's familiar *Almagest*. Consequently, astronomers, while wholly aware of the multiform defects of Ptolemaic astronomy, retained the use of those calculating techniques until something better was forthcoming: how reminiscent this is of nineteenth century astronomers' retention of Newtonian mechanics despite its complete failure in accounting for the aberrations in the perihelion of Mercury. In *De Ludo Globi* Cusa suggests that permanent rotation is a property of perfect sphericity. Thus, were a billiard ball a perfect sphere, its twisting movement would continue *ad indefinitum*. God had only to give the outermost celestial sphere its original *impetus*, and, by its perfect sphericity, it has continued to rotate ever since and to keep all other mechanically-subordinate celestial spheres in motion. This hypothesis actually influenced Copernicus, as we shall see.

Cusa's contemporaries, Georg Puerbach (d. 1461) and Regiomontanus (i.e. Johannes Müller, d. 1476) are usually referred to as the immediate predecessors of Nicholas Copernicus (b. 1473). Puerbach, a practical observational astronomer well within the Ptolemaic tradition, helped to revive the Alfonsine Tables, using sines rather than chords. Regiomontanus computed a table of sines for every minute, and a table of tangents for every degree; he completed a text book started by Puerbach; it was called the *Epitome in Ptolemaei Almagestum* (Venice, 1496). Regiomontanus' pupil, Bernard Walther (d. 1504) used a clock driven by a hanging weight for the timing of astronomical observations at the observatory in Nuremberg.

Although these men may have waxed cosmological during their informal and non-professional conversations, we remember them today exclusively for their devotion to accuracy and detail in the observational and calculational study of the heavens. Naturally, they invoked the most precise and helpful techniques in plotting their descriptions and predictions. These were uniformly Ptolemaic, but by this time articulated with rather more powerful mathematical techniques than were available during the second century A.D.

To encapsulate this pre-Copernican situation in the late fourteenth and fifteenth centuries, let us quote from an illuminating paper by Dr. Derek Price:

[Before Copernicus] ... the two parts of astronomy were so detached, one from the other, that they formed almost separate, mutually exclusive subjects having little bearing on each other ... On the one hand was the subject of descriptive and physical cosmology, predominantly Aristotelian, with its objective of understanding the heavenly universe and relating it to the rest of man's knowledge ... On the other hand was the highly complex mathematical theory concerned with the phenomenological determination and prediction of the irregularities of planetary motions ... There never was such a thing ... as a Ptolemaic system.[32]

This captures a good deal of what we have been trying to delineate. In very ancient Greece astronomy and cosmology began life as one undifferentiated *study of the heavens*. Eudoxos' superior mathematical skills, and Plato's comprehensive philosophical outlook, however, already began to bifurcate this study. From the Hellenistic period to the high Middle Ages these disciplines remained distinct, although, as we have seen, they again become somewhat interfused in the rigorous logical inquiries of the fourteenth and fifteenth centuries. One way of estimating Copernicus' stature in the history of planetary theory is to recognize that he was at once enough of a philosopher and enough of a mathematician to seek a single set of integrated *explicantia* which might individually afford (1) at least as much precision in description and prediction as was available in the Ptolemaic techniques, and (2) also dovetail into a philosophically-satisfying cosmological explanation of the heavens. The success of this mighty undertaking we must characterize in the next section of this book.

---

[32] In *Contra–Copernicus* in 'Critical Problems in the History of Science', ed. Clagett, p. 199.

# BOOK TWO

## PART II

Copernicus' Systematic Astronomy

In 1453 Constantinople fell to the Turks. But this event had already been built into the two preceding centuries. The year 1543 has more than a numerological connection with the fall of an ancient empire; for in that year the *De Revolutionibus Orbium Coelestium* was published, on the very day Copernicus died. The Western mind has never been the same since; it is almost as if when the medieval curtain fell, Copernicus went to join his predecessors of the previous two centuries, whose critical explorations find their consummate achievement in his *magnum opus*.

We have already seen that in Cusa's *De Docta Ignorantia* the universe is conceived to be infinite in extent; it is therefore devoid of any center whatever, and of any circumference. It follows that the earth cannot be in the center of the universe. Moreover, since motion is natural to all bodies, the earth itself must be in motion. Cusa notes that a person standing at the North Pole of the earth, and another standing at the North Pole of the celestial sphere, would *both* think themselves to be in the center of the universe. How like today's Cosmological Principle! The world is a wheel within a wheel, a sphere within a sphere, with nowhere a center or a circumference.[1] As with Buridan, Cusa remarks that motion is detectable only by comparison with fixed objects; in mid-ocean the sailor may think his ship stationary, though it be making twenty knots. Cusa did not treat the Sun as a celestial object in the traditional sense, i.e., as a perfect quintessential entity. Within the solar body there is a *quasi terram centraleorem* – the apparent fire resides only at the circumference. And, in the famous note by Cusa, in his own handwriting[2] it is denied that any motion can be exactly circular. No stars describe exact circles from rising to setting; no point within the stellar sphere can be a permanent pole. The earth moves like the other stars. Only, where the eighth sphere revolves from East to West in twelve hours, the Earth does so in twenty-four hours. This makes a terrestrial observer think the earth immovable

[1] *De Docta Ignorantia*, Liber II, Ch. IX, Paris 1514.
[2] Discovered by Clemens, cf. *Giordano Bruno und Nicolaus Von Cusa*, Bonn 1847.

while the stars revolve each twenty-four hours. In this note Cusa makes clear his debt to Eudoxos, for he enspheres one sphere after another and adjust the poles and speeds in a very subtle way.[3] But the corruptibility of the heavens, and the eccentric mobility of the earth, these distinguish Cusa as an independent thinker of the first rank.

Peurbach, the younger contemporary of Cusa, was quickly becoming a technically proficient Ptolemaic astronomer. His text book *Theoricae Novae Planetarum* became a much-reprinted foundation for all later astronomical commentary. It is one of the clearest and most concise expositions of the Ptolemaic technique. But Peurbach was not satisfied only to generate the consequences of arbitrary hypotheses, in the style of a fourteenth or fifteenth century logician. Although he was fundament-ally a predictional astronomer, he aspired at least to that degree of under-standing advertised in then-contemporary cosmologies. Cusa also mani-fested this trait. Georg Puerbach fitted the solid crystalline spheres which the Arabs had inherited from the later Aristotelians solidly into his astronomy. But he did not pack them tightly together; he left room between each of the spheres to allow free play to the excentric orbit and epicycle required for each planet. He became increasingly desirous of working through Ptolemy's *Syntaxis Mathematica* in the original Greek, as contrasted with the third hand Latin-through-Arabic versions then available. A friend, Cardinal Bessarion, was also anxious to make the Greek literature better known in Northern Europe. Due to Puerbach's death (1461 A.D.) his pupil Regiomontanus journeyed south in his stead to commence a study of Greek under the Cardinal's tutelege.

Regiomontanus was, of course, a native of Königsberg. After many years with Bessarion, Johannes Müller (that was his real name) returned to Nürmberg and compiled the astronomical *Ephemerides* which were of such great help to the Portugese, Spanish and Genoese navigators. In his *Tabulae Directionum* he included a table of sines for every 1′ and a table of tangents for every 1°. His fame was so considerable that the Pope summoned him to Rome in 1475 to reform the Calendar. But, like his teacher Peurbach, Regiomontanus also died before he was forty. But not before completing the important text book begun by Peurbach *Epitome in Ptolemaei Almagestum* (Venice 1496). In that tract

---

[3] *Op. cit.*, D 287.

Regiomontanus advocates the Ptolemaic technique in every minute detail. Because of Schoener's publication (1533) of a tract entitled *An Terra Moveatur an Quiescat Joannis De Monte Regio Disputatio* many historians of science credited Müller with being a precursor of Copernicus himself.[4] But Dreyer completely disproves this.[5] Doubtless Regiomontanus carefully re-articulated every feature of the planetary analyses originated by Ptolemy. Regiomontanus is important because, what with his powerful mathematical accomplishments, his command of Greek and a tested capacity to read all ancient texts, his prowess as an observational astronomer and his considerable gifts as a lecturer and a teacher – he is virtually the fifteenth century reincarnation of Claudius Ptolemy himself. He is a more redoubtable champion for the cause that was smitten in 1543, than had been the great Constantinople just before 1453.

Leonardo da Vinci is very often cited as another anticipator of Copernicus. Libri and Venturi, among many others, have advanced this view. But even if this were true – and it has certainly not been established – Leonardo's work was not *known* in detail anywhere north of the Mediterranean continent.

Celio Calcagnini (d. 1541 A.D.) did for certain teach the daily rotation of the earth. Some time before 1545 he wrote *Quod Coaelum Stet, Terra Moveatur, Uel de Perenni Motu Terrae* (*Opera Aliquot Basilieae*, 1544). Therein Calcagnini denies that the heavens, Sun and Stars, are rocketing around the Earth each day; it is the Earth which revolves. In another eight pages Calcagnini appeals to a mass of allusions concerning the way the earth is like a flower – always turning its face towards the Sun – that it is in the center, twisting diurnally, as contrasted with the light and pure quintessential celestial bodies which are immobile. Calcagnini does note that the earth's rotational motion is not perpetual; it inclines from side to side, as is shown in the solstices and equinoxes, the increase and decrease in the size of the moon, the varying lengths of shadows, and so forth. He notes also the obliquity of the ecliptic, how the moon recedes as much as 5° from the Zodiac, the trepidation of the stellar sphere and so forth – all as constituting further complications within the diurnal rotation. Calcagnini even notes that since Archimedes

---

[4] Doppelmayr, Veidler, Montucla, and Schubert are among these.

[5] *A History of Astronomy*, 290–91.

promised to move the earth if he had a place to stand on, the earth must therefore be movable! He quotes Cicero re: Hiketas and Plato's *Timaeus*, as well as alluding to Cusa himself. All of which shows that Calcagnini was aware that other people had taught the rotation of the earth. But, alas, he attempts to account for every conceivable celestial perturbation by this *one* rotational hypothesis.

In a publication that is fascinating chronologically, Francesco Maurolico undertakes to refute the 'absurd opinion' of Calcagnini, and others, that the earth moves.[6] In every way Maurolico's tract is almost purely medieval. The Sun's orbit is placed midst the planetary orbits for the simple reason that the inferior and superior planets are quite different concerning the periods of their epicycles and deferents.[7] Through all his long life Maurolico was a loud anti-heliocentrist. His attack on Calcagnini in 1543 shows how lively was the controversy over the explanation of the 'second inequalities' of planetary motion, into which Copernicus then injected the *De Revolutionibus*.

Through all this time a great number of books on *The Sphere*, patterned after the singular treatise of Sacrobosco, were printed and reprinted right on through to the sixteenth century. Alexandrian trigonometry and spherical geometry, as well as Ptolemaic astronomy, therefore, were very well known in learned circles in the early sixteenth century; the original *Syntaxis Mathematica* was printed at Basle in 1538 from a codex brought north by Regiomontanus. But the desire to 'cosmologize' the Ptolemaic computer into something that might make the universe intelligible was waxing ever more strongly.

G. Fracastoro knew Copernicus in 1501 at Padua. In 1538 he published *Homocentrica* in which he undertakes to revive the theory of solid spheres, converting Ptolemy's calculator into a *physically true system* of the universe. He dedicated the work to Pope Paul III, to whom Copernicus also dedicated *De Revolutionibus* five years later. The salient idea in Fracastoro's treatise is that *all* the Eudoxian-Callipian crystalline spheres should have their axes set at right angles to each other. His idea seems to be that, inasmuch as any spatial translation can be resolved into three components at right angles to each other, any celestial motion

---

[6] Cf. *Cosmographia*, Venus, 1543.

[7] The solar year is the period of the deferent for the inferior planets, and the period for the epicycle in the superior planets.

should therefore be analysable in these same terms. Fracastoro is motivated by the same 'mechanical' objectives that animated Aristotle's cosmology. But, in order to embrace all the aberrations and perturbations which observation of the heavens had made known by this time, he requires an intricately-complex cluster of spheres. These spheroidal complications led to a sphere cluster seventy-seven in number; this is eleven more than the sixty-six we urged here for Aristotle (Book One, Part I).

In 1536 G. B. Amici published a treatise on the homocentric system, and this quite independently of Fracastoro. It is much more clearly written than Fracastoro's work and treats the problems of general planetary perturbations in a much more general way. He does not, as with Aristotle, require one sphere α to provide for each planet-cluster a diurnal rotation. After reviewing the theories of Aristotle, Eudoxos and Kalippus, Amici proclaims directly that nature does not contain such things as epicycles and excentrics. Again, all of the observational details (by then well-known) made Amici's cosmology, like Fracastoro's before him, much more complex than Aristotle's. But the driving idea behind all these was the same – a desire for an intelligible cosmological *system*. The arbitrary mathematical inventions – excentrics, deferents, and epicycles – so well known to precise, analytical astronomers of the day, could not satisfy such men as these, men who wished not only accurately to describe and predict, but also to know why their descriptions and predictions turned out to be accurate – nature being what it was. Just as Aristotle could not rest content with the chaotic toolbox of computers passed along to him by Eudoxos and Kalippus, so also Calcagnini, Fracastoro, and Amici – following through fields furrowed by the criticisms of Grosseteste, Scotus, Ockham, Cusa, Buridan and Oresme – could not easily be satisfied with the astronomical computing tools which had found their way into the sixteenth century all the way from Ptolemy's second-century workshop.

Everything seemed now ready. Many were emboldened to criticize the bifurcation between prediction and explanation, between astronomy and cosmology, a criticism absent from Western thought almost since the time of Aristotle. The habit of criticizing tradition was by now becoming well engrained, however. The mathematical learning and skill necessary to make a significant advance of any kind was now at hand.

All one needed was that single individual with the courage to criticize accepted traditions – with the acumen to articulate in detail every fresh insight – with a deep desire to be something more than merely a celestial engineer, a 'mere predicter'; what was needed was a *philosopher of the heavens* who really sought to understand the structure of the universe. And along came Copernicus.

Like Burkhardt's *homo universale*, Nicolas Copernicus was master of so much that he could refuse the slight calculational advantages offered by the technical astronomy current amongst his contemporaries in the early sixteenth century. From his philosophical studies in Italy he knew what a rational understanding of a perplexity was like. His medical studies had made clear the value of systematizing wherever possible. And in his legal studies he saw again and again the values of clarity and economy in exposition. His prowess as a mathematician has been universally recognized, despite the faint dissent of Derek Price.

In this Act, History wrote the perfect plot: the actor possessed of the greatest gifts and skills was also he who, for broader intellectual reasons, could not and would not be satisfied with prediction *simpliciter*, nor with explanation *simpliciter*. Copernicus' dissatisfaction with the astronomies and cosmologies of two millenia of predecessors was like the dissatisfaction a 16th century Hempelian would have felt. These accounts were all woefully incomplete. Either they made description and prediction impossible because of too-hasty cosmological postures, or else they pursued the *minutiae* of analytical astronomy rendering the cosmos *en toto* thereby incomprehensible.

Copernicus' master stroke, of course, was his heliocentric reconfiguration of the planetary system for the purpose of accounting for the so-called 'second inequalities' – the positions of the planets at conjunction and opposition, as well as their stationary points and retrograde arcs. The basic idea is to us quite simple. It had not been overwhelmingly difficult for Aristarchus, and even Ptolemy continually acknowledges the formal advantages of a moving earth. (See Figure 50.) One immediately apparent feature of this scheme is its interconnexity. The several positions of the planets at any one time are simultaneously representable in such a Copernican configuration. This is in marked contrast to the representation characteristic of the Ptolemaic astronomy; as we saw many times before (see Figure 51a). A complete *lack* of

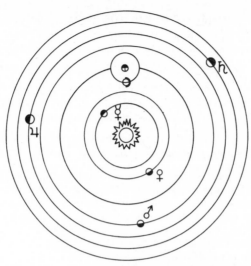

Fig. 50.   Copernicus' universe. The Copernican system as described in *De Revolutionibus Orbium Coelestium*. N.B. This is the first workable astronomy which is also representable as a cosmological picture.*

interconnexity is what identifies the computing devices illustrated in Figure 51a. But inasmuch as stationary points, and retrograde arcs, are not (for Copernicus) special problems to be dealt with *seriatim* by constellations of epicycles, deferents and equants – but rather fundamental structural features of the physical geometry of the planetary system itself – this forces any one calculation within Copernicus' theory to involve (implicitly) every other possible planetary calculation. Indeed, this is a dramatic anticipation of the later Laplacean dictum that given a complete understanding of the state of the universe at any one moment, a mathematically 'possessed' demi-god could calculate every other cosmical configuration past and present. The 'systematic' character of Copernicus' centric relocation did, ideally, involve the astronomer in a wholesale cosmological commitment – not simply in a piecemeal nibbling at planetary perturbations. Or, to express it in another way: Ptolemaic calculations were always specific solutions to particular two-body problems-some planet and the earth. Copernicus' calculations, however, must always spring from multi-body considerations – at least in principle.

* *Editor's note:* But see Editor's note on p. 5, *supra*, concerning Ptolemy's system.

For first he must orientate the earth with respect to the sun, the system's center. Then the planet in question must be orientated *both* with respect to the sun (for computation) and with respect to the earth (for observation). But then the observed positions of the other planets are immediately relevant as well. Because there is no explanatory virtue in accurately describing the planetary configuration during one observation at time *t* if at some future time *t'* Mars seems to be 'right on course', while Venus and Jupiter are worlds away from where they were expected to be. As he says himself in his *Dedication to the Pope*:

... I have, assuming the motions which I in the following work attribute to the earth, after long and careful investigation, finally found that when the motions of the other planets are referred to the circulation of the earth and are computed for the revolution of each star, not only do the phenomena necessarily follow therefrom, but the order and magnitude of the stars and all their orbs and the heaven itself are *so connected* that in no part can anything be transposed without confusion to the rest and to the whole universe. (My italics.)

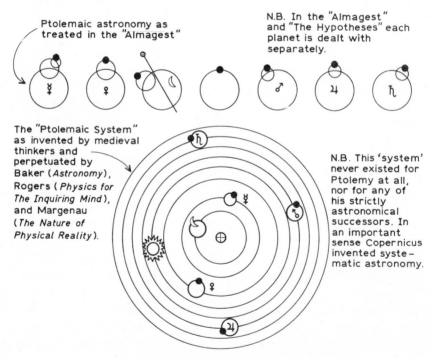

Ptolemaic astronomy as treated in the "Almagest"

N.B. In the "Almagest" and "The Hypotheses" each planet is dealt with separately.

The "Ptolemaic System" as invented by medieval thinkers and perpetuated by Baker (*Astronomy*), Rogers (*Physics for The Inquiring Mind*), and Margenau (*The Nature of Physical Reality*).

N.B. This 'system' never existed for Ptolemy at all, nor for any of his strictly astronomical successors. In an important sense Copernicus invented syste-matic astronomy.

Fig. 51a.   Ptolemy's general scheme of calculation (Halma).

Copernicus' Conception of Retrograde Motion

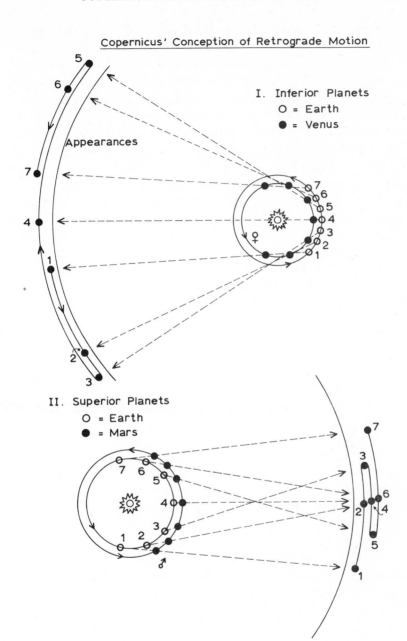

Like other close students of the stars before him, Copernicus was struck by the fact that the revolution of the sun through the zodiac and the revolution of the epicycle-centers of Mercury and Venus through the zodiac both took the same amount of time – one year. He noted also that in geocentric astronomy the epicyclical period of the three outer planets, Mars, Jupiter and Saturn, was identical to the synodic period.[8] As we have pointed out, this sun-dependent character of the planetary motions had been noted by many astronomers and cosmologists. But, for Copernicus, it was suggested that perhaps the deferents of the two inner planets and the epicycles of the three outer ones represented simply an orbit passed over by the earth in one year – and not by the sun at all! *Apparentias salvare*; but at this stage Copernicus was only the Aristarchus of the Renaissance – no more. To articulate and implement this insight with the sharp instruments of technical mathematical astronomy, *this* was the Copernican achievement. In this much alone Aristarchus and Copernicus both succeeded in relating the fact that Mercury and Venus are tethered to the sun with the further fact that when Mars, Jupiter and Saturn are in opposition, their motion is retrograde.

This dazzling insight consumed Copernicus at some stage. He seems to have wished to invest in the earth as many motions as were necessary to account for the planetary phenomena – such as the duration of Venus' (epicycle-center's) 'orbit' around the earth, and the necessity of giving the outer planets a period of one year in their epicycles. For he says that the earth's motion must be such "as to appear outside (similiter extrinsecus) in many ways, from which we recognize the annual revolution."[9a]

The relativistic equivalence of alternative descriptions of motion is clear from this. Just as Buridan had noted the observer's freedom to describe himself as in a moving boat within a stationary sea, or as within a stationary boat set in a moving sea – so here Copernicus notices that if certain motions must be attributed to the planets in order to achieve *apparentias salvare*, the same observational effect would follow from investing the earth with the geometric reciprocal of those motions. Thus either Orion turns around us each day, or we twist before Orion's face once each day. When opposite the sun (from our point of view) either Mars 'moves backwards', or we, on the 'inside track', move past

[8] The time between two successive oppositions to the sun.
[9a] *De Revolutionibus*, Liber I, Ch. 9.

it (as it is seem against the fixed stars). These alternative descriptions are observationally equivalent, in a sense to be examined. The decision to opt for one, rather than the other, is made, in all cases, on the basis of extra-kinematical considerations – e.g., cosmological ones. Thus, although it is of great importance to remark that Copernicus' major stroke was to advocate a new cosmological commitment, it is false to say (with Dr. Price) that he was therefore something less than a brilliant mathematician. Before the Copernican system and the Ptolemaic techniques could be put into an equipose requiring scientific decisions, the former had to be sharpened to the same degree of descriptive precision and predictive accuracy as had been achieved by the former. Price points out that *De Revolutionibus* was no more accurate than the *Almagest*, from which he seems to conclude that it did not really constitute a major mathematical advance. What should be stressed is that *only* a mathematical advance of the first order could achieve for Copernicus' cosmological insights the descriptive and predictive articulation available then in the sixteenth century to the Ptolemaic technique. Copernicus' fate ultimately has been quite different from that formerly known by the ancient Aristarchus. The major difference was the monumental mathematical machinery which turned the computational wheels within *De Revolutionibus*.

Many of these wheels were patterned along the design-lines of the Mercurial and Venusian theory described by Martianus Capella – the so-called 'Egyptian System'. Thus:

And if one takes occasion of this to refer Saturn, Jupiter and Mars to *the same* center, remembering the great extent of their orbits which surround not only those two but also the earth, he will not miss *the explanation* of the regular order of their motions.[9b]

In a reference to Philolaus and Aristarchus, Copernicus says:

... As these things are such as cannot be understood except by a sharp mind and prolonged diligence, it remained at that time hidden to most philosophers, and there were but few who grasped *the reason* of the motions of the stars ....[10]

It is a genuine feature of Copernicus' outlook that he seeks an explanation, an understanding, of the totality of sidereal motions – not only of the phenomena involving revolutions of the planets, but also those

[9b] Lib. I, Ch. 10.
[10] My italics.

encountered when the latter sometimes are seen to move away from us, sometimes to approach us,[11] "which do not agree with the assumption of concentricity." It should be clear that the expression *apparentias salvare* takes on a different semantical force with each new use, by each new speaker. For, in Copernicus, this expression almost never means *merely* 'saving the appearances' in the sense of generating particular phenomenological descriptions of phenomena, whatever the cost to physical intelligibility. Very often Copernicus uses these words in reference to the observational consequences of a theory *en toto*. When this is so, his language is much like that of a contemporary hypothetico-deductive theorist. He argues against the physical truth of a theory, or a hypothesis, whenever some of its important consequences fail to square with the facts. This is quite different from a much more limited use of *apparentias salvare* in which the logical plausibility of a formal computer is established *merely* from the fact that descriptions of phenomena tumble out at the bottom of the page – whatever might be the state of one's understanding of the physical correlates of the predicting devices. This anticipates the issue between Schlick's 'verifiability' criterion and Popper's 'falsifiability' criterion. The merely Ptolemaic version of *apparentias salvare* seems very like what has been known as the 'black box' conception of a physical theory. Copernicus was almost certainly driving for a 'glass box' conception of a physical theory – one which told the truth in detail, because it transparently depicted phenomena truly, and in general. Contrast this with some more recent views of a physical theory as being nothing more than an algebraic box into which observation-numbers are poured, the inference-handle turned, with descriptions and predictions tumbling out at the bottom. This latter may be close to what some of Copernicus' Ptolemaic predecessors had in mind as constituting a satisfactory astronomy. It is clearly different from what he had in mind – and different too, I suspect, from what Carl Hempel proposed as the optimum relationship between explanation and prediction.

Long before the publication of *De Revolutionibus*, Copernicus, at the request of friends and colleagues, produced a *Commentariolus* (a short summary) – a sketch of his planetary system; it was circulated

---

[11] E.g., Venus' variations in brightness.

in manuscript. In a short Introduction it refers to the failure of Eudoxos' theory to account for the varying distances of the planets (i.e., their variable brightness); the highly objectionable character of the equants of Ptolemy are also alluded to. It was these latter, indeed, with their *ad hoc* arbitrariness, which constituted the primary motivation for Copernicus to seek some new arrangement of planetary circles. In seven subsequent chapters Copernicus then sets out the order of the orbits, the triple motion of the earth (rotation, revolution, and the conical 'wobble' of the spinning terrestrial axis). He then stresses the advantages of linking all celestial motions not with the terrestrial equinoxes, but with the fixed stars themselves. Next, he deals with the proposed circular constructions for the lunar motions, those for the outer planets, and for Venus and Mercury. But he discusses only the *relative* sizes of the orbital circles – no proofs or detailed reasoning of any kind is given. Indeed, only a reader with some fairly advanced acquaintance with the Ptolemaic techniques could hope to follow the exposition of the *Commentariolus*. This treatise is, in many ways, typically medieval; there is no stress in it on the *physical* truth of the hypothesis of the earth's mobility. This is, rather, treated as but one more hypothesis from the assumption of which one can achieve *apparentias salvare* – in the limited 'black box' sense. As was pointed out by Birkenmajer (Warsaw, 1933), the *Commentariolus* is not simply a finished summary of the contents of *De Revolutionibus*. Rather, it constitutes but an earlier stage in the development of Copernicus' mature planetary theory. The former is 'concentrobiepicyclic' – the later work is 'eccentrepicyclic'; in the former case the planet was located on the circumference of the second epicycle [12] which itself moved uniformly around its moving center – this being situated on the circumference of a first epicycle revolving uniformly about a moving center fixed on the deferent. [13] In *De Revolutionibus*, however, Copernicus went over to eccentrepicyclism, wherein the planetary epicycle is made to ride directly on an eccentric deferent. [14] Copernicus was well aware of the equivalence, illustrated earlier in Figures 28 and 29, between the eccentric and epicyclical arrangements.

---

[12] the 'epicyclepicyclus' (*DR* 218.29).

[13] Copernicus' lunar theory (*DR* 235.14–236.8) is an example of concentrobiepicyclism, as well as the entire planetary theory set out in the *Commentariolus*.

[14] *DR* 325.16–326.31.

They yielded the same results. This is clear in Ptolemy's *Almagest* (Heiberg, Vol. I) 216.18-217.6 – with which Copernicus was intimately familiar. But, by the sixteenth century, the use of eccentrics had been complicated by the introduction of the equant (Cf. Figure 45). Copernicus regarded this device as totally improper, and rejected the *punctum equans* as being completely 'unphysical'. For him the *ratio absoluti motus* required every epicyclical circle to move not only uniformly but also uniformly with respect to its own center. His brilliant disciple Rheticus put it this way:

An essential property of circular motion is that all circles in the universe should revolve uniformly and regularly about their own centers, and not about other centers.[15]

And, as Copernicus himself puts it in *De Revolutionibus*:

For when [the ancients] assert that the motion of the center of the epicycle is uniform with respect to the center of the earth, they must also admit that the motion is not uniform on the eccentric circle which it describes ... But if you say that the motion uniform with respect to the center of the earth satisfies the *ratio absoluti motus*, what sort of uniformity will this be which holds true for a circle on which the motion does not occur since it occurs on the eccentric?[16]

In any event, the *Commentariolus* is a stage in the development of Copernicus' thinking toward the *De Revolutionibus*. From epicyclic to eccentric calculations.

Referring again to Mars, Jupiter and Saturn (I, 10) Copernicus observes that these planets are nearest to the earth when in opposition to the sun (the earth being then between them and the sun), but that they are farthest from us at superior conjunction (when the sun is between them and the earth). Thus, clearly, their center belongs in the sun – that around which Venus and Mercury themselves obviously move.

Therefore we are not ashamed to maintain that all that is beneath the moon, with the center of the earth, describe among the other planets a great orbit around the sun which is the center of the world; and that what appears to be a motion of the sun is in truth a motion of the earth; but that the size of the world is go great that the distance of the earth from the sun though appreciable in comparison to the orbits of the other planets, is as nothing when compared to the sphere of the fixed stars.[17]

[15] From Prowe, *Nicolaus Coppernicus* (Berlin, 1884) Volume II, 318-28-31.
[16] *DR* 233.11-13, 233.29-234.1.
[17] Note here exactly the argument Archimedes had conveyed to us as being that of Aristarchus.

He continues:

I hold it easier to conceive this than to let the mind be distracted by an endless multitude of circles, which those are obliged to entertain who detain the earth in the center of the world. The wisdom of nature is such that *it produces nothing superfluous* or useless, but often produces many effects from one cause. (My italics.)

Note here a ringing echo of Scotus and Ockham.

We shall make all this clearer than the sun, at least to those who know something of mathematics. The first principle is that the size of the orbits is measured by the period of revolution, and the order of the spheres ... commences with the uppermost.

Notice the qualitative anticipation here of Kepler's third law − $T^2 \propto r^3$ − linked with the further notion that the ordering of the planets is part of the systematic structure which relates radial distance and angular velocity.[18]

Copernicus continues:

The first and highest sphere is that of the fixed stars, ... immovable, being the place in the universe to which the motions and places of all other stars are referred. For while some think that it also changes somewhat [N.B., a reference to precession: N.H.] we shall, when deducing the motion of the earth, assign another cause for this phenomenon. Next follows the first planet, Saturn, which completes its circuit in just under thirty years then Jupiter, with a twelve years period [7862 years, actually] then Mars, which moves round in two years [1881 years, actually]. The fourth place in the order is that of the annual revolution, in which we have said that the earth is contained with the lunar orbit as an epicycle. [We must never forget that these vestiges of epicyclical astronomy have a prominent role in Copernicus; they are not extinct in Newton, and even in ourselves today]. In the fifth place Venus goes round in nine months [224701 days, actually]. Finally Mercury, with a circuit of eighty eight days [87969 days, actually].

[18]  These things are, of course, totally unconnected in the Ptolemaic technique. Remarks about planetary order in the *Almagest*, and all the machinery of.the epicycles, eccentrics and equants − as well as observations of celestial angular velocity, variable planetary brightness and arcs of retrograde motion − these are in no way linked within the calculational techniques employed. The planets could be ordered in any way whatever; radial distance determinations are independent of the main technique. The planets could be moving at infinity − the extant calculations still applying equally well. Such remarks as are made about order and distance are forced upon the Ptolemaic astronomer by cosmological considerations merely. They were not determined by the technique as a whole. Here the Copernican 'system' differs profoundly. All these things are interrelated in the *De Revolutionibus* like elements of a Chinese puzzle. Change e.g., the numerical value of a planetary distance-variable, and you immediately necessitate far-reaching readjustments throughout the entire planetary complex. All other elementary variables for all other planets immediately require overhauling: Copernicus' system was in principle like a Swiss clock − interconnected at every couple and joint. Such interconnexity is totally foreign to the Ptolemaic technique.

But in the midst of all stands the sun. For who could in this most beautiful temple place this lamp in another or better place than that which can at the same time illumine the whole? Which sun is not unsuitably called the light of the world, or the soul or the ruler. ... So indeed the sun, sitting on the world throne, stirs the revolving family of stars.

Many writers, and most especially Kuhn[19] have remarked the neo-Platonic sentiment in this passage, with its virtual sun worship. Indeed, Marsilio Ficino might have written it himself, as he did write:

Nothing reveals the nature of the Good [which is God] more fully then the light [of the sun] ... the sun can signify God himself to you ...[20]

Even Kepler chooses to express his adulation for the newly discovered importance of the sun thus:

the sun "is a fountain of light, rich in fruitful heat, most fair, limpid, and pure to the sight ... and which alone we should judge worthy of the Most High God ... we return to the sun ... worthy to become the home of God himself, not to say the First Mover."[21]

Doubtless, there is a great deal to this insight, and Kuhn uses it well. But Copernicus was a man of many parts; *homo universale*. It is for the historian of astronomy to stress those features in the total Copernican complex which seem most to illuminate his particular conceptual objectives. That Copernicus thought well of the sun is indisputable. But perhaps one should not make too much of this secondary fact. Neo-Platonic sun-worship itself could not reasonably be established as *the* motivation behind *De Revolutionibus*. Copernicus' ardent desire for a fully unified, systematic and wholly intelligible explanation of our local celestial configuration – mathematically articulated down to the finest particular – it is within the framework of this desire that the sun (for Copernicus) comes to assume its greatest importance.

Copernicus concludes this section of his treatise by forging another link for the systematic reticulum of his heavenly study. The very geometry he ascribes to the solar system immediately makes clear *why* the retrograde arc of Jupiter is greater than Saturn's and smaller than that of Mars – and why Venus' is greater than that of Mercury. He says a few more words concerning why the outer planets are brightest when in opposition – all these things being simply the observational and optical effects of terrestrial motion.

---

[19] *The Copernican Revolution*, pp. 128ff.
[20] *Liber De Sole*, in *Opere* (Basel, 1576).
[21] From Burtt, *The Metaphysical Foundations of Modern Science* (1932), p. 48.

Concerning the third motion attributable to the earth – that required by the 'backward' movement of the points of intersection of ecliptic and equator, the 'precession of the equinoxes' – Copernicus reveals himself as having been a close student of the works of Eudoxos. He even subjects the earth's axis to two further motions, perpendicular to each other, the combined effect of which is to move the pole of the earth through the circumference of a lemniscate, along the points *ABCDEFGHI*. (Figure 51 b.) Not only is this path described as a result of Eudoxian composition – a primitive form of harmonic analysis – but we have here a direct reflection of the 'hippopede' set out in Figure 18.

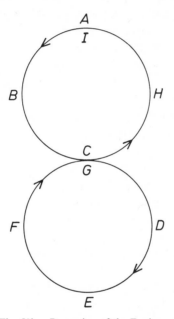

Fig. 51b.   Precession of the Equinoxes.

But this again demonstrates Copernicus' central strategy – which is to investigate the consequences of attributing every stellar aberration to some new motion of the earth. Given the clarity of his recognition of the geometrical relativity of observations of motion, and given further his commitments to the fixity of the firmament and the *ratio absoluti motus*, one could imagine that *any* new observation of an unexpected aberration in stellar motion would immediately be an indication, for Copernicus, that some new degree of freedom had to be added to the total account of terrestrial movement. At this stage, indeed, Copernicus does seem to be sacrificing somewhat the principle of perfectly circular motion. It is here that he demonstrates that any motion whatever can be shown to result from *some* combination of circular motion; in this case he generates rectilinear movement.[22] Nasir ed-din Al Tûsi had used this technique in his own planetary theory. Kepler, as has been suggested, also used it in his theory of Mars – but it was almost certainly known to Hipparchus and Apollonios.

By taking the annual orbit geocentrically-attributable to the sun, and placing it in the earth, Copernicus was able simply to account for the 'second inequalities' – the retrograde planetary motions, the inconstant points of opposition, the varying brightness of Venus, the orders, mean distances and mean periods of all the planets ... etc. This he did in one architectonic revision which, by itself, made a cathedral out of a pile of stones (to exploit Poincaré's metaphor). But this monumental, systematic simplicity was not quite enough. For, as we know, retroflectively, the 'problem of the empty focus' would eat away at any planetary construction which failed to take account of it.

The considerable variations in the angular velocities of the planets as they traversed their orbits ultimately required modifications of the two major commitments Copernicus had already made, and with which he had successfully challenged all then-extant astronomical techniques. These were the Principle of Perfect Circularity (of which much shall be made in a moment) and the Principle of Absolutely Uniform Motion along a deferent. Only the recognition of the ellipticity of the planetary 'orbits', combined with the Law of Central Forces, could realign these 'First Inequalities'. Indeed, it may have been the very recognition of these observational inequalities which forced Copernicus' predecessors indirectly to tinker with the Principle of Perfect Circularity (as in the epicycles, eccentrics and equants) and with the unqualified equidistance of all planets' orbital positions from the earth. Which is another way of indicating how important the systematic aspects of *De Revolutionibus* actually must have been to its author. By paying attention to the Second Inequalities a systematic interconnexity of all planetary perturbations

[22] Compare above Figure 31.

was achievable – despite a far less satisfactory treatment of the First Inequalities. Copernicus' predecessors were so concerned with these latter inequalities – the First – that they seemed prepared to sacrifice all 'systematics' for the advantages of accuracy in description and prediction.

Copernicus' choice was, historically, the correct one. For he achieved intra-astronomically what his predecessors could achieve only by an appeal to cosmology, namely, the linking and reticulating of the Second Inequalities. This step was indispensable to any fully satisfactory treatment of the First Inequalities (by way of Kepler's First Law, and the Law of Central Forces).

Nonetheless, let us never imagine that Copernicus' predecessors were taking the 'easy way out'. In a sense, they chose the more difficult type of problem, and did whatever they could, even to the sacrifice of Eudoxian simplicity (which Copernicus restored) in order to do justice to the phenomena.

Concerning fine-structured descriptions of the earth's orbit around the sun, Copernicus added virtually nothing to the eccentrics which Ptolemy had used to cope with the solar motion. He simply translated the original eccentricity of the orbit, and the longitude of the apogee in the sun's travel to the aphelion in that of the earth. But, because of the commitment to circularity, and considerable errors in ancient observations, he had to configure the sun and the earth as in Figure 52. Here, the sun is at $S$, the earth at $E$. The center of the terrestrial orbit, $B$, rolls around the point $A$ from east to west once every 3434 years, while $A$ moves from west to east around the sun in 53000 years. In this attempt to account for the motion of the aphelial point, the 'problem of the empty focus' is pellucidly clear. By this arrangement if the radius of the terrestrial orbit $BE$ equals 1, then $SA$ equals 0.0368 and $AB$ equals 0.0047. In the arrangement as in Figure 53, the eccentricity is at a maximum – a phenomenon which actually took place in 64 B.C. But as set out in Figure 52 the eccentricity is minimal and the aphelial point is moving fastest. Because of the relocation of the planetary system's center in *De Revolutionibus* Copernicus' lunar theory works out to be much simpler then that of Ptolemy. A single epicycle gives him the equation of the center. But where elsewhere his planetary calculations are 'eccentrepicyclic', here in his lunar theory he simply mounts a second epicycle on the first – rendering the computations thus 'concentrobi-

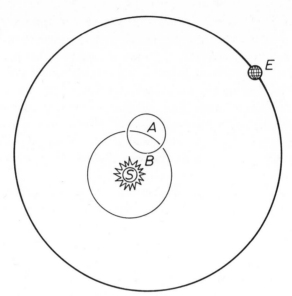

Fig. 52.   Sun and earth in Copernicus.

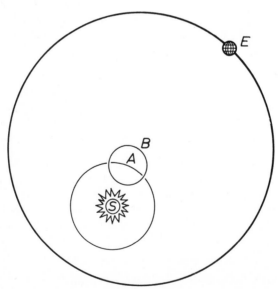

Fig. 53.   Maximum eccentricity of earth's circumsolar path.

epicyclic'. The deferential center is thus at $E$, the center of the earth; on its circumference the center of the first epicycle moves from west to east with the mean sidereal motion of the moon. The second epicycle moves on the circumference of the first in the opposite direction with the mean anomalistic motion ($13°3'53''56''.5$ per day). The moon, $M$, moves on the second epicycle from west to east twice in every lunation; it is at $F$ at every mean syzygy and at $M$ at every mean quadrature.[23] Thus the enormous changes in parallax resulting from the Ptolemaic technique are avoided by Copernicus.[24] The greatest distance of the moon from the earth Copernicus sets at $68\frac{1}{3}$ semi-diameters of the earth; the smallest distance was $52\frac{17}{60}$. The apparent diameter of the moon thus varies between $28'45''$ and $37'34''$, an enormous improvement on the theory of Ptolemy.[25]

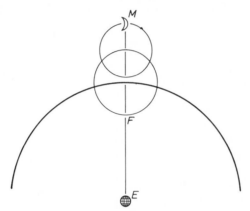

Fig. 54.   Copernicus' lunar theory.

In his individual planetary calculations, Copernicus also had the advantage over his Ptolemaic predecessors. For the Second Inequalities which had given so much additional trouble to Ptolemy now fall into place as an architectural feature of the planetary system; Copernicus has thus only to attend to the First Inequalities – each planet's sidereal period of revolution. For this Ptolemy had used eccentric *and then*

[23]  Lib. IV, Ch. 3.
[24]  In Ptolemy's systems, remember, the diameter of the moon at perigee would be almost $1°$, which is clearly counter-observational.
[25]  IV, Ch. 17.

equant. But, as a matter of principle, Copernicus could not make that appeal. In Figure 55a $S$ is the center of the earth's orbit. The center of each planet's eccentric orbit is at $C$; now the planet, $P$, moves on an epicycle in the same direction and with double the angular velocity with which a center moves around the eccentric. The radius $AP$ is one-third the eccentricity $CS$ of the deferent[26].

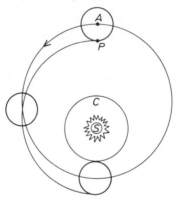

Fig. 55a.   Copernicus' general planetary computation.

This arrangement, referred to as the 'eccentrepicyclum', is generally preferred by Copernicus to the *epicycli-epicyclium*.

Inasmuch as Mercury presents problems near the end of this book at least as great as Mars will present in the immediate sequel, it seems appropriate to point out that, due to the great eccentricity of this planet, Copernicus took the problems it posed very seriously. Its eccentricity is twice as great as that of Mars; hence the simple arrangement Copernicus had used for Venus will not work for that planet at all. (See Figure 55b.) Here, whenever the earth is at $E$ or $E'$ the center of the small circle is set at $N$. Moreover the planet does move on the eccentric, but backwards and forwards on the line $KL$.[27] This line is always directed toward the center of the eccentric. So, Mercury will be at $K$ every six months, whenever the earth is at $E$ or $E'$ and the center of the

[26] Indeed $CS$ plus $AP$ is equal to Ptolemy's eccentricity of the equant.
[27] The diameter of a small epicycle, what Kepler later calls "librati in diametro epicycli"; in *De Motibus Stellae Martis*.

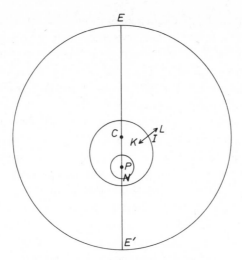

Fig. 55b.   Copernicus' Mercurial plot.

eccentric at *N*. As Copernicus says:

Therefore Mercury, by its proper motion, does not always describe the same circle, but very different ones according to the distance from the center, the smallest when at *K*, the greatest when at *L*, the average when at *I* [numerically, the greatest circle equals 0.3953, the smallest equals 0.3573, the mean equals 0.3763, *DN* equals 0.0212, *CD* equals 0.0736 (Ch. 27)] ... what in the case of the moon is done in the circumference, that occurs at Mercury in the diameter by two reciprocal motions, composed of equal ones.[28] And how this is done we have already seen when dealing with the precession of the equinoxes.[29]

Here then is another subtly cloaked deviation from the principle of strict circular motion. It must be obvious enough that this is something Copernicus would do only when extreme complexities of the phenomena require it. Mercury is certainly a case in point – being a tormentor for the greatest astronomers right up into the twentieth century.

Indeed, Copernicus' Mercurial and Venusian techniques are every bit as intricate as anything found within the Ptolemaic corpus. For both Mercury and Venus the line of the nodes[30] coincides with the line of apsides.[31] Hence the greatest departures in latitude from the ecliptic ought to take place when the planets are 90° from aphelian. However,

---

[28] I suppose this means uniform circular, but counter-rotating, motions: N.H.
[29] Book V, Ch. 25.
[30] Where the orbital-plane and the ecliptic-plane intersect.
[31] Connecting the perihelial and aphelial points.

two kinds of 'librations' affect their motions. The first has a period of one-half year; whenever the sun passes through the perigee or apogee of the planet, the inclination of the orbital plane is greatest. The other libration takes place around a moving axis through which the planet always passes whenever the earth is 90° from the apsides; but when the apogee or perigee of the planet is turned towards the earth, Venus always deviates most to the North, and Mercury most to the South. If the sun falls in the apogee of Venus, and the planet happens to be in that place as well, then the simple inclination and first libration produce no latitude, but the second one which takes place around an axis at right angles to the line of apsides produces the greatest deviation. But had Venus then been 90° from the apsides, the axis of this libration would pass through the (mean) sun and Venus would add to the northern 'reflection' the greatest 'deviation', or decrease the southern by the same amount. (See Figure 55c.) If *abcd* is the earth's orbit and *flgk* the eccentric

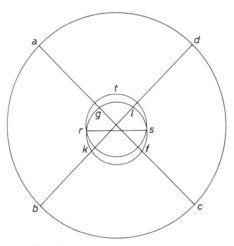

Fig. 55c.   Copernicus' Venusian plot.

orbit of Venus or Mercury at its mean inclination to the former, *fg* will be the line of the nodes. When both earth and the planet are in the line *ac* the planet has no latitude, but it will have if it is in either of the two semi-circles *gkf* or *flg* – and that latitude is called 'obliquation' or 'reflection'. But when the earth is at *b* or *d*, these latitudes in *gkf* or *flg* are called 'declinations,' "... they differ from the former in name rather than in reality." [32] But since the inclination is found greater in the obliquation than in the declination, this is taken to be caused by another libration around *fg* as its axis. A 'circle of deviation' is assumed, which is inclined to *gkfl*, concentric with it for Venus but eccentric to it in the case of Mercury. Their line of intersection *rs* is a moving axis of this libration. When the earth is at *a* or *b* the planet will reach its limit of deviation at *t*; and while the earth moves away from *a* the planet moves from *t* at the same rate – while the circle's obliquity decreases. When the earth has reached *b* the planet has reached the node *r* of this latitude. But then the two planes coincide and change positions, so that the other semicircle of deviation (hitherto to the south) will now go to the north, and Venus (before in the north) will continue there and will never be turned to the south. Similarly Mercury will deviate *only* to the south, and, naturally, the period for both planets is just one year.

All this constitutes but the slightest modification to Ptolemy's own theory. Ptolemy had greater difficulty in representing the Mercurial and Venusian latitudes because he had let the line of nodes pass through the earth rather than through the sun. The full discovery of this had to await Kepler – and it was a discovery which affected the entire development of planetary theory thereafter. Copernicus also erred in this respect, by letting the line of nodes pass through the center of the earth's orbit which, even for this redoubtable heliocentrist, was not identical with the sun. This displaced the nodes so that Mercury or Venus was bound to have some latitude when it ought to have had none – being placed squarely in the ecliptic at that time. The degree of this erroneous latitude varied with the place the earth happened to be in its orbit – which is in itself a clue to where the line of nodes *should* have been placed. It was all this that made some oscillations in the orbits necessary, even oscillations of a markedly rectilinear variety.

[32] VI Ch. 2.

# FURTHER ASPECTS OF COPERNICAN ASTRONOMY
## IN CONTRAST TO ALL THAT HAD GONE BEFORE

Professors A. R. Hall, D. J. de Solla Price, T. Kuhn, M. B. Hall and J. L. E. Dreyer have seriously misdescribed the 'equivalence' of geostatic and heliostatic astronomy, as well as that between the Copernican and Tychonic systems. Nothing less than our full understanding of the development of planetary theory is at stake here. Hence a critical analysis of the claims and intentions of these distinguished historians may be tolerable.

In a penetrating account of the mathematical relations between the Ptolemaic techniques and the Copernican construction, Professor A. R. Hall writes:

The *geometrical equivalence* of the geostatic and heliostatic methods of representing the apparent motions of the celestial bodies, adopted by Ptolemy and Copernicus respectively, is not often clearly emphasized though it is vital to any discussion of the nature of the change in astronomical thought effected by Copernicus.[33]

Professor D. J. de S. Price argues:

It follows from this principle [of geometrical relativity] that the use in Ptolemaic theory of a geostatic deferent with epicycle is *strictly equivalent* to a heliostatic system in which the epicycle, transferred to a central position, becomes a second 'orbit'.[34]

Although it is correct to insist that the geostatic-geocentric and heliostatic-heliocentric techniques are equivalent in some sense, both Hall and Price (and Kuhn and Dreyer, as we shall see) err in characterizing the nature of that equivalence. It will be argued here that the linking relationship is certainly not one of 'geometrical equivalence' (Hall and Kuhn); nor are the two techniques 'strictly equivalent' (Price). They are most assuredly not 'absolutely identical' (Dreyer); not in any sense of 'absolute identity' or 'strict equivalence' familiar to logicians of science. A more correct description would be that the two approaches are 'observationally equivalent', in a sense to be made clear.

---

[33] *The Scientific Revolution*, p. 370 (my italics).
[34] 'Contra-Copernicus', in Clagett: *Critical Problems in the History of Science*, p. 203 (my italics).

Hall and Price compress their arguments beautifully into the following demonstration: consider an arbitrary Ptolemaic construction for either Mars or Venus (see Figures 56a and 56b).

In such a geostatic representation the geometrical-kinematical principles invoked will be the same whether the planet is Superior like Mars, or Inferior like Venus. This will not be so in a heliostatic construction, where different constructions are necessary for the Inferior planets (cf. Figures 10(a) and 10(b)).

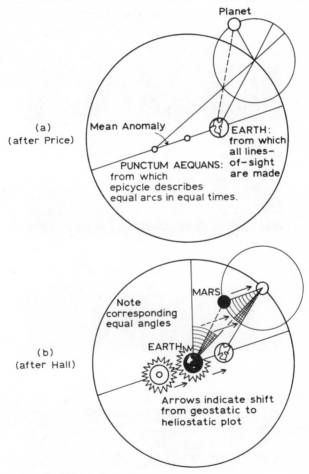

Fig. 56a–b.   Generalized Ptolemaic planetary plot.

Now, what *are* the observations with respect to which 'apparentias salvare' is achieved by such a Ptolemaic construction? They are as set out next in Figure 57. In Figures 58, 59, 60 and 61 immediately following, the 'standard' (and simplified) geostatic and heliostatic techniques for the 'saving' of these appearances are combined in a manner which I have not seen elsewhere.

As is well known, if the planet happens to be Mars, *apparentias salvare* can be achieved in detail by use of the generalized geostatic construction in Figure 59, as follows in Figure 62; the thick, arrowed path is the resultant, geometrically-generated, Martian orbit for one year.

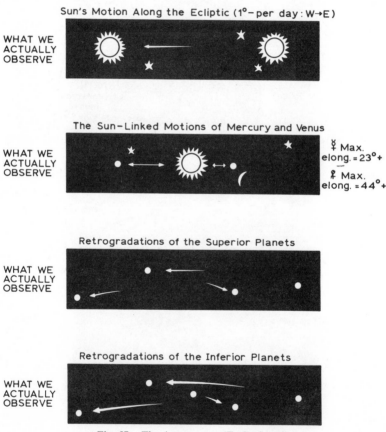

Fig. 57.   The Appearances 'To Be Saved'.

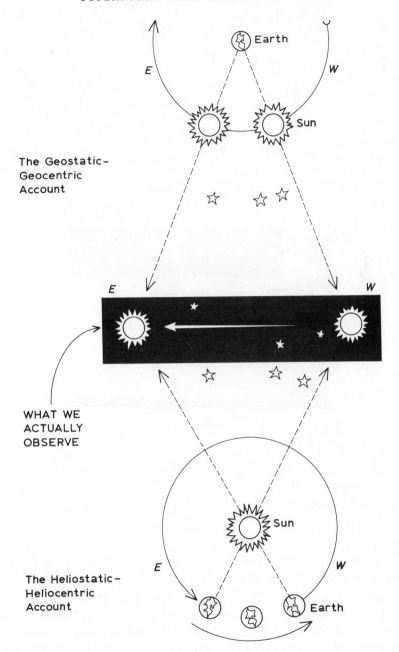

The Geostatic-
Geocentric
Account

WHAT WE
ACTUALLY
OBSERVE

The Heliostatic-
Heliocentric
Account

Fig. 58.   The sun's ecliptic motion.

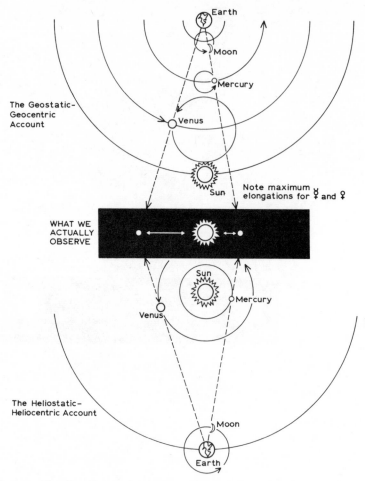

Fig. 59.   The sun-dependent motion of Mercury and Venus.

The generalized heliostatic-Copernican representation for Superior planetary motion (corresponding to Figure 58) is next set out (Figure 63a), followed by its detailed specification for Mars (Figure 63b).

As both Hall and Price point out, there is considerable geometrical isomorphism between Figures 62 and 63b. The ratio of epicycle radius to deferent radius in Figure 62 in principle exactly equals the ratio of ter-

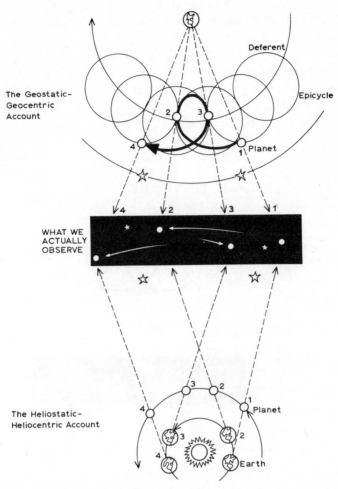

Fig. 60. Retrogradations of the superior planets.

restrial-orbit radius to Martian-orbit radius in Figure 63b.[35] And the angle Mars–Earth–epicycle center in Figure 62 exactly equals the angle Sun–Mars–Earth in Figure 63b.

If, however, the planet is Inferior, e.g., Venus, then, although the general geostatic arrangement will remain as in Figure 1, the general heliostatic configuration is different (Figure 65a), and the specific helio-

[35] Actually, this latter was shortened somewhat to fit it on the page.

static treatment of Venus is as in Figure 65b. The specific geostatic treatment of Venus is as in Figure 64.

The degree of isomorphism here between Figures 64 and 65b is identical to that between Figures 62 and 63b. The geometrical ratios all remain constant (although the earth's orbit has been again shrunk slightly for representational purposes).

What Hall and Price seem to have in mind is this: there will always in

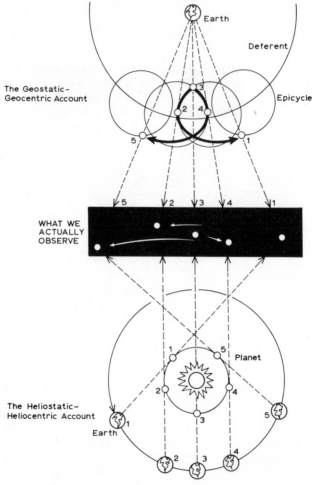

Fig. 61.   Retrogradations of the inferior planets.

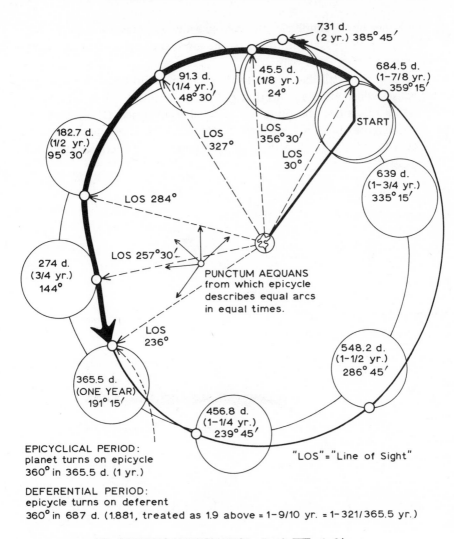

731 d.
(2 yr.) 385° 45′

684.5 d.
(1-7/8 yr.)
359° 15′

91.3 d.
(1/4 yr.)
48° 30′

45.5 d.
(1/8 yr.)
24°

LOS
327°

LOS
356° 30′

LOS
30°

START

182.7 d.
(1/2 yr.)
95° 30′

639 d.
(1-3/4 yr.)
335° 15′

LOS 284°

LOS 257° 30′

274 d.
(3/4 yr.)
144°

PUNCTUM AEQUANS
from which epicycle
describes equal arcs
in equal times.

LOS
236°

548.2 d.
(1-1/2 yr.)
286° 45′

365.5 d.
(ONE YEAR)
191° 15′

456.8 d.
(1-1/4 yr.)
239° 45′

EPICYCLICAL PERIOD:
planet turns on epicycle
360° in 365.5 d. (1 yr.)

"LOS"="Line of Sight"

DEFERENTIAL PERIOD:
epicycle turns on deferent
360° in 687 d. (1.881, treated as 1.9 above = 1–9/10 yr. = 1–321/365.5 yr.)

(Cf. SYNTAXIS MATHEMATICA – Book XIII, 1–6)

Fig. 62.   Ptolemaic circumterrestrial computation for Mars. (After Figure 56.)

principle be some 'Translation Rule' with which one can move from a
geostatic representation such as those in Figure 56, 62, or 64, to a
corresponding representation such as those in Figure 63a, 63b, 65a or
65b. It is clear that all the observations in Figure 57 can be 'accounted

for' either by the simplified geostatic or heliostatic constructions set out in Figures 58, 59, 60 and 61, or by the more detailed configurations in Figures 62, 63b, 64 or 65b.

This is really all that Hall and Price claim in those sections of their treatises which are our concern here.

Does this much constitute 'geometrical equivalence' (Hall) between what is represented in Figures 56, 62, 64 and what is set out in Figures 63a, 63b, 65a, 65b? Are the two representations 'strictly equivalent' (Price)?

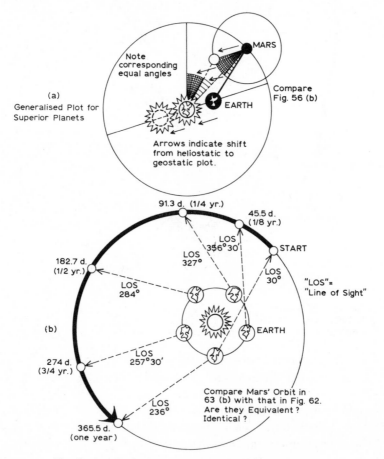

Fig. 63a–b.   Copernican circumsolar computation for Mars.

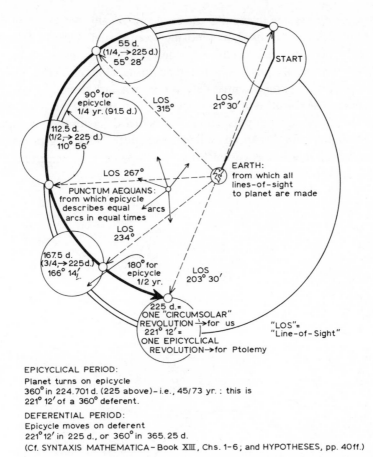

Fig. 64. Ptolemaic circumterrestrial computation for Venus. (After Figure 56.)

They are not. In any standard sense of *geometrical equivalence*, or strict equivalence, one theory $\theta_1$ will be equivalent with another theory $\theta_2$ only if one can infer all of $\theta_2$ from $\theta_1$ and all of $\theta_1$ from $\theta_2$. $\theta_1 = \theta_2$ only when $\theta_1$ and $\theta_2$ are *mutually entailing*. Thus, for example, an epicyclical representation of any arbitrary planetary orbit will be geometrically/mathematically/strictly/formally/logically equivalent to any corresponding eccentric representation of that orbit. This is made clear both in the *Syntaxis Mathematica* (III, 3) and in *De Revolutionibus Orbium Coelestium* (Thorn ed., 207.2–3). Everything geometrically generable by the one technique, can be geometrically generated by the other as well. A point fixed on a moving eccentric deferent will describe precisely the *same orbit* in physical space as it would were it set on an epicycle (of radius equal to the deferent's original eccentricity), whose center moves on a non-eccentric deferent. If, in a dark room, the 'planetary' point stood out as a phosphorescent dot and described an oval path, a viewer could not possibly know whether this path was generated by an eccentric or by an epicyclical mechanism. *That* is geometrical equivalence! It was recognized even by Hipparchus as constituting a

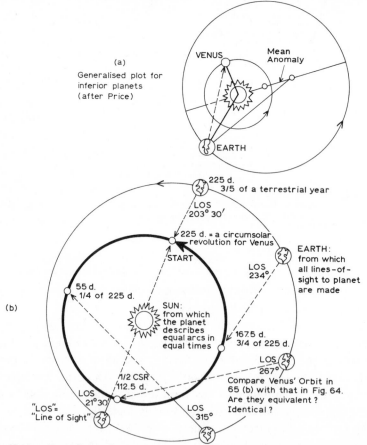

Fig. 65a–b. Copernican circumsolar computation for Venus. (*De Revolutionibus*, Book V, Ch. 22.)

different *kind* of equivalence from that with which Hall and Price are concerned.

A more obvious case: if $\theta_1$ consisted in the claim:

> *the Moon describes a circumterrestrial orbit every point on which is equidistant from a given point;*

and $\theta_2$ reads:

> *the Moon describes a circumterrestrial orbit which encloses a maximum area for that length perimeter* –

one must by the same criterion grant strict, geometrical equivalence of $\theta_1$ and $\theta_2$. The curve generated by $\theta_1$ will, in physical space, be congruent with that generated in physical space by $\theta_2$. In other words, $\theta_1$ and $\theta_2$ have exactly *the same geometrical consequences*. Hence they are equivalent in the strictest, most formal sense.

Refer again now to Figures 62, 63b, 64 and 65b. Most emphatically, these constructions do *not* generate the same resultant orbital curves in physical space for Mars and Venus. Indeed, *there* is the mathematical heart of the differences between Ptolemy's collection of calculational techniques, and Copernicus' systematic astronomy. Because, while the computations of *De Revolutionibus* could be given a direct physical interpretation, those of the *Almagest* could not. The dark orbits in Figures 63b and 65b were construable as configurations in physical space. But those in Figures 62 and 64 were not: they were merely the abstract loci of selected points corresponding to *lines of sight* from earth to planet. Clearly, there will be an infinite number of possible orbit-shapes compatible with such lines-of-sight. This is not true of the Copernican design. Those in 62 vs. 63b, and in 64 vs. 65b, are clearly dissimilar. Yet they are the geometrical products of $\theta_1$, the Ptolemaic technique, and $\theta_2$, the Copernican. Therefore, $\theta_1$ and $\theta_2$ are *not* geometrically equivalent, or strictly equivalent, Price and Hall notwithstanding. Because it is not the case that everything generated from $\theta_2$ is indifferently generable from $\theta_1$. It would be difficult to understand what the Copernican revolution was a revolution *about*, were this not the case. For if $\theta_1$ and $\theta_2$ *were* geometrically equivalent, in the strict sense, there could be *no* distinguishing the consequences deducible from them. There could be *no* different expectations from $\theta_1$ and $\theta_2$ concerning the phases of Venus, the configurations of retrograde loops *as seen from* Polaris ... etc. But there certainly were historically significant differences *vis-à-vis* such expectations. Therefore $\theta_1$ and $\theta_2$ are demonstrably not equivalent in any strict sense.

In the Twentieth century it has been argued (by Schrödinger and Eckart) that Wave Mechanics and Matrix Mechanics were strictly and mathematically equivalent in just the sense under discussion. What *that* assertion amounted to was that every consequence deducible from the one $\theta$ was also deducible from the other, and vice versa, without remainder. This was explicitly stated as *the meaning* of the claim.[36] The equivalence claim as between Wave Mechanics and Matrix Mechanics in the Twentieth century must be characteristically different from the equivalence claim, tendered by Hall and Price, as it purports to obtain between the geostatic and heliostatic theories of Sixteenth century astronomy.

But, although definitely not formally equivalent, the Ptolemaic technique and the Copernican construction are equivalent in some sense. What sense? In just that sense historians of science seem long to have recognized, and which it was the purpose of the 'LOS''s, in Figures 62, 63b, 64 and 65b to delineate in detail. Both techniques $\theta_1$ and $\theta_2$ succeed in 'saving the appearances' or, in more modern terms 'squaring with the observed phenomena': the Lines of Sight in Figure 62 and 63b and in Figures 64 and 65b are identical! But no historian of science should argue from this to the *formal equivalence* of the two theories in question. No one would seek to infer from the fact that, during the Eighteenth century, both the Corpuscular Theory of Light and the Undulatory Theory of Light squared with the observed phenomena – that therefore the two theories were internally and formally equivalent. As we all know, from a formal and internal point of view they were very dissimilar indeed. This, despite the fact that many computational techniques (e.g., the generation of specific refractive indices from certain applications of Snell's Law), are common to both these optical theories. Indeed, *from* the fact that a deduction involving Snell's Law and some refractive index obtained in the Corpuscular Theory, one could infer that it would also obtain in the Undulatory Theory. After all, both theories were trying to square with the known facts. And both were *succeeding* in the

---

[36] This assertion was challenged by the present writer in *Current Issues in the Philosophy of Science*, ed. Feigl and Maxwell, Holt, 1961; and in his book *The Concept of the Positron* (Cambridge University Press, Autumn, 1962).

Eighteenth century. In the Sixteenth century both $\theta_1$ and $\theta_2$ were having a comparable success. But formal equivalence is no more a property of the latter pair than of the former – 'apparentias salvare' notwithstanding.

As Hall and Price point out, there are many geometrical analogies and techniques which, if used in the description of planetary positions, and articulated according to a Ptolemaic technique, will ensure a corresponding description of the positions (observed angles, angular velocities, points of opposition and conjunction, etc.) in the Copernican construction. Should one expect anything else? The whole point of this geometrical-astronomical game was to generate 'apparentias salvare', so clearly some isomorphism should be expected as one sets out his initial conditions in Ptolemaic, or Copernican, terms. *Something* common to both sets of initial conditions, and premises, must be such as to entail the actually observed planetary positions (i.e., the Lines of Sight!). Both of these astronomies succeeded, in the Sixteenth century, in doing that. So it is not unreasonable to suspect assumptions and techniques common to both approaches.

Professor Derek Price begins by seeking to "establish the fundamental principle of geometrical similarity which shows that angles, distances, and indeed all mathematical techniques are quite unaffected by any change from geostatic to heliostatic systems."[37] Price concludes that the two systems are "strictly equivalent" – a conclusion to which the foregoing constitutes a challenge. The argument Price sandwiches between these dissimilar slices of premise and conclusion is as follows:

The case illustrated is indeed only a special case of a more powerful theorem which might be used to consider three points in space, S(un), E(arth), and P(lanet), all moving in any paths whatsoever. If one considers S to be at rest, the ends E and P of vectors SE and SP will trace out paths which we may call the orbits of E and P respectively. Now, if one takes E as being at rest instead, S will appear to move in an orbit marked out by the vector ES. Since this is exactly the negative of SE, the orbit will be geometrically identical but turned through 180°. In this 'geostatic' system the orbit of P will now be given by the motion of the sum of the vectors ES and SP, the former having been just discussed, the latter now becoming an 'epicycle' added to this orbit. It should be noted that this argument is perfectly general and not restricted to a plane; it shows that planetary latitudes as well as longitudes are included in the principle of geometrical relativity between geostatic and heliostatic systems. Provided that suitable lines and planes of reference are maintained there is not a single mathematical technique or calculation which is peculiar to the geostatic or the heliostatic system alone.[38]

[37] *Op. cit.*, p. 203.
[38] *Op. cit.*, p. 203.

The earlier part of Price's argument is quite sound, and constitutes a general formula for the guarantee that two astronomical theories will be 'observationally equivalent' – which it was the object of Figures 62 through 65 to delineate. Price's conclusion, however, does not follow: there certainly *are* calculations which are peculiar to the geostatic or to the heliostatic system alone – namely, the generation of those resultant orbits for Mars and Venus set out earlier. 'Lines and planes of reference *are* maintained' in generating these non-equivalent orbits. At this point a more general issue emerges.

The orbit geostatically generated in Figure 62 is the result of but one leading consideration. In order that the Lines of Sight from Earth to Mars should be calculable-as-observed, an epicycle and a deferent of *any* arbitrarily suitable dimensions and angular velocities are chosen. The orbit which results is entirely controlled by the observations 'to be saved', and by the quite arbitrary choices of epicycle-deferent ratios, the size of the equant, etc. The orbit for Mars on the heliostatic scheme 63b does not have its 'shape' determined in this way. Rather, it is determined by 'systematic' considerations, indeed cosmological considerations – those which enter the *Almagest* only as independent afterthoughts. Copernicus has *two* things to save: the appearances (of course) and the 'system'. Ptolemy sought salvation only for the first. Copernicus' eyes require 'apparentias salvare' just as did Ptolemy's; his mind, however, requires some intelligible constellation of configurations such that the appearances turn out to be precisely what one would have expected, the geometry of the planetary system being what it is.

Price's 'relativistic' formula is a general guarantee of whether certain $\theta$s save the appearances. Thus, if a theory $\theta_1$ 'saves the appearances' and another theory, $\theta_2$, is geometrically related to $\theta_1$ as per Price's formula – then it follows that $\theta_2$ will also save the appearances. The Lines of Sight generable in both will be identical. There can be no exception to this. This, indeed, is a *sufficient condition* for saying of any $\theta_2$ (so related to a $\theta_1$ that *is* observationally confirmed), that it *also* achieves 'apparentias salvare'. It is obviously not a *necessary condition*: there could be other theories, $\theta_n$, which also 'save the appearances' (by widely different calculational techniques), but which do not comply with the formula. Thus, any $\theta_2$ which is related to any $\theta_1$ *via* this geometrical relativity formula, will be observationally equivalent to $\theta_1$. But it is *not necessary* that every

$\theta$ observationally equivalent to $\theta_1$ be also related to it *via* this formula. The class of observationally equivalent theory-pairs is much wider than the class of theory-pairs linked *via* the geometrical relativity formula.

This formula is not a sufficient condition for establishing the *geometrical* equivalence between $\theta_1$ and $\theta_2$, to put an earlier point back-to-front. Because, there could be a $\theta_1$ and a $\theta_2$ related *via* this geometrical-relativity formula – for example, geostatic and heliostatic astronomy – which are nonetheless quite different in that they generate resultant geometrical orbits which are wholly dissimilar. On the other hand, any two $\theta$s which are geometrically equivalent must necessarily be related *via* the geometrical relativity formula. It is thus a serious conceptual error to construe 'the formula' as constituting a *necessary and sufficient* condition for geometrical equivalence between $\theta_1$ and $\theta_2$, as Price, Hall, and others apparently do: it is *only* a sufficient condition for establishing observational equivalence between $\theta_1$ and $\theta_2$. Or, it is a necessary condition for establishing geometrical equivalence between $\theta_1$ and $\theta_2$. It is not both. This is the logical kernel which itself encapsulates the illustrated argument set out before.

*What is no more than a necessary condition for establishing the geometrical equivalence between two theories, or is no more than a sufficient condition for establishing the observational equivalence between two theories, has been construed by Hall and Price as a necessary and sufficient condition for establishing their formal equivalence.*

Simply generating the same conclusions is wholly insufficient (as we saw) to guarantee the equivalence of two theories. $\theta_1$ and $\theta_2$ could be different in every conceivable way – logically, notationally, sequentially and psychologically – and *still* generate the same observational consequences.

Hall and Price know well that very dissimilar theories, lacking any kind of translation techniques permitting inference from the one to the other, might nonetheless generate the same observational consequences. Theories at opposite logical poles may yet 'save the appearances'. Part of the intention of Hall and Price is to indicate that the Ptolemaic technique and the Copernican constructions are *not* really all that much different from a logical-geometrical point of view. These theories have many things in common. *Why* this should be so has been a matter for illuminating exploration in recent studies in the history of science – to which Hall and Price have eloquently contributed. There are the best contextual, historical and conceptual reasons for having *expected* Copernicus to have used much that was already available in the Ptolemaic treatises. Nonetheless, Hall and Price have etched this point in unwarranted manner. Let us agree that the geometrical similarities between the Ptolemaic techniques and the Copernican constructions are *much* more striking and pervasive than has been generally recognized until recently. Let us agree that both approaches indifferently 'save the appearances'. At least for 'naked-eye' astronomy. These two points of agreement, however, do not warrant the conclusion of both Hall and Price, to wit; that the Ptolemaic techniques and the Copernican constructions were therefore formally equivalent.

Someone will interject here that the resultant planetary orbits within the *Almagest* were never meant to be mapped into physical space, and that hence to compare their geometry with those of *De Revolutionibus*, which were so planned, is illegitimate. What then of Brahe? Were his 'orbits' meant to picture reality, or were they mere computers too?

When discussing the structural relationship between Copernican and Tychonic astronomy, Professor Kuhn refers to their 'geometrical equivalence', (*The Copernican Revolution*, page 204.) He even says that "Mathematically the only possible difference between the motions in the two systems is a parallactic motion of the stars ..." (*op. cit., loc. cit.*).

A. R. Hall and M. B. Hall also describe the two systems as 'mathematically equivalent' (in 'Tycho Brahe's System of the World', *Occasional Notes of the Royal Astronomical Society*, Vol. 3, No. 21, (1959), page 254.)

Indeed, the great J. L. E. Dreyer says of these two planetary systems that they are "absolutely identical" (*A History of Astronomy*, Dover (1953), p. 363).

As with the Ptolemaic and Copernican orbital extrapolations in Figures 62, 63b, 64 and 65b, the following illustration is the only one I know of a full plot for the Martian orbit, according to the Tychonic technique. Again, as with Figure 62 and 63b, the *Lines of Sight* here generated are identical with those of the Ptolemaic and Copernican Martian plots. But the resultant orbit – a geometrical entity geometrically generated – is wholly unlike those which issue from the other two constructions. (See Figures 66–68.)

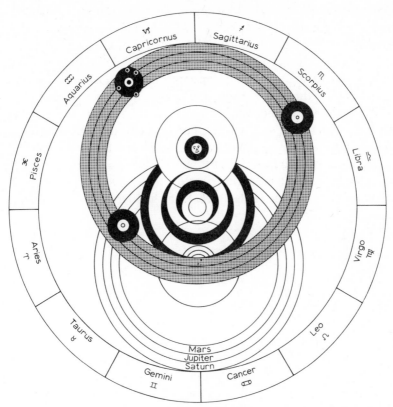

Fig. 66.   Tycho's system.

The conclusion is just as before. What Kuhn, Hall and Dreyer rightly mean to stress is that all Lines of Sight generated in this Martian problem *are the same*, irrespective of whether the plotting technique is Ptolemaic, Copernican or Tychonic. They do this misleadingly by characterizing all three as being 'geometrically equivalent', 'mathematically equivalent', 'strictly equivalent', and 'absolutely identical'. Saying such things does not permit us to distinguish the kind of equivalence binding the epicyclical and eccentric techniques (or that of our hypothetical lunar theories $\theta_1$ and $\theta_2$; or that claimed for Wave Mechanics and Matrix Mechanics in 1926), from *this* situation wherein the different planetary $\theta$s issue in different resultant orbits *ad indefinitum*. The equivalence before us is no more than an observational equivalence; it is considerably reinforced (as the Undulatory-Corpuscular observational equivalence is not) by the fact that this kind of equivalence is guaranteed to obtain between the Ptolemaic, Copernican and Tychonic constructions in virtue of the "principle of geometrical relativity" which links them (cf. Price, *op. cit.*, p. 203). But all this means is that, from the *Line-of-Sight*-viewpoint these three theories stand or fall together: the LOSs in Figures 62, 63b and 68 are, and will always be, identical. That situation will obtain with any three $\theta$s which are thus geometrically equivalent. Line-of-Sight Equivalence *via* geometrical relativity is a necessary condition of the geometrical equivalence of $\theta_1$ and $\theta_2$. But just because this LOS equivalence does obtain, as with *our* $\theta_1$, $\theta_2$ (and $\theta_3$, the Tychonic) it does not follow – as the Halls, Price, Kuhn and Dreyer assert – that the

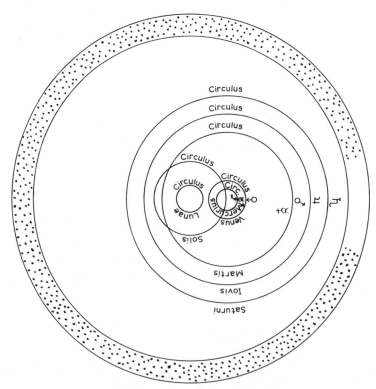

Fig. 67.  Tycho's system by Hevelius.

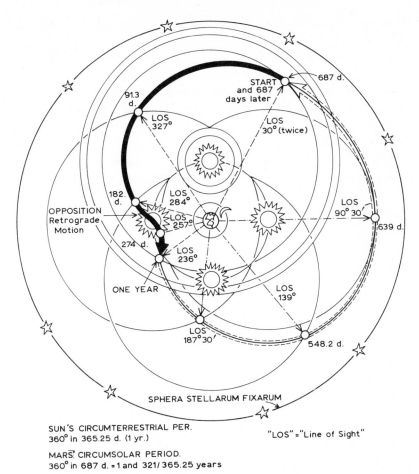

Fig. 68.   Tychonic 'Circumterrestrial-Circumsolar' computation for Mars (from *De Mundi Aetherei Recentioribus Phaenomenis*, Ch. VIII).

Ptolemaic, Copernican, and Tychonic techniques are geometrically equivalent. The former is not a sufficient condition of the latter.

> *When theories are strictly equivalent it follows that they are observationally equivalent. But when they are observationally equivalent, even because bound by the "principle of geometrical relativity", it does not follow that they are geometrically, mathematically, or strictly equivalent.*

Confusion on this point of scientific logic is enough to distort the entire history of planetary theory, and of the individual contributions of the key figures within that history. One might even seek to minimize the mathematical prowess of Copernicus because of the claim that his computations were mathematically equivalent to those of Ptolemy.[39] But the claim is logically false. So the appraisal must rest on other grounds.

From all the foregoing it should be clear in what sense the Ptolemaic technique and the Copernican construction were equivalent. They were not equivalent in any thoroughgoing *formal* sense. But they did use a considerable number of mathematical devices in common, *but organized quite differently* – and they rendered up the same predictions of what would be observed were an astronomer at a particular time to cast his gaze in a particular celestial direction. This 'difference in organization' is precisely the systematic innovation for which the history of western thought shall always be specifically indebted to Copernicus.

Before discussing in much more general terms Copernicus' mathematical prowess and his relative stature as contrasted with that of Ptolemy before him, and Kepler and Newton after him, let us first discuss Copernicus' entire enterprise somewhat more discursively.

We have stressed already the 'systematic' character of Copernicus' exposition. The Aristotelian cosmology he had inherited from his late medieval teachers was groping and blind. The Ptolemaic astronomy of which he had become the technical master was arbitrarily empty. The former did not permit him to say anything in articulate detail about very much at all. The latter gave him no insight whatever into the physical *reasons* why the celestial kinematics to which he and his contemporaries were phenomenologically exposed manifested themselves as they actually did, rather than in a host of easily-imagined alternative ways. In Copernicus we experience the terminus of the exclusive dichotomy which had dominated the entire history of celestial studies before his time; the dichotomy between empty, directionless, non-systematic predictive astronomy – as against groping, inarticulate, wooly, explanatory cosmology.

---

[39] This is Price's thesis in *Contra-Copernicus*.

The *De Revolutionibus* was, of course, very closely modeled on the exposition of the *Almagest*. Indeed, we find in Copernicus the first display of technical competence required to take on Ptolemy's problems point for point as they actually occur in the *Syntaxis Mathematica*: for the first time in almost 1400 years a figure here strides onto the astronomical stage who could have sat down with Ptolemy in the second century A.D. and wrangled over matters geometrical and predictional. This is in no way to underestimate all of the serious cosmological and astronomical reflections during the intervening period. But that Copernicus' intellectual relationship with Ptolemy can be described in the foregoing way at all has always been one of the main reasons for the rather unilluminating historical characterizations of the intervening centuries as being 'dark' or 'medieval'. The actual astronomical observations during the interim were negligible. It is almost as if Copernicus has been locked away with Ptolemy's treatises for over a millennium, then to be released to take up precisely the same problems as did *The Astronomer* – with an eye now toward more mathematical intelligibility, and systematic elegance. Referring to the Ptolemaists, Copernicus says:

With them it is as though an artist were to gather the hands, feet, head and other members for his images from diverse models, each part excellently drawn, but not related to a single body, and since they in no way match each other, the result would be a monster rather than a man. So in the course of their exposition, which the [Ptolemaic] mathematicians call their system ... we find that they have either omitted some indispensable detail or introduced something foreign and wholly irrelevant.[40]

As Kuhn points out[41] the astronomical stage onto which Copernicus strode displayed not only Ptolemy's *Almagest*, but a host of other theoretical starlets as well; multiform and myriad Islamic and European astronomies, although they criticized and modified and decorated the original Ptolemaic technique by adding, or deleting, epicycles or eccentrics, or by bringing new measurements or ratios into the original system – nonetheless, all these geostatic computors were fundamentally Ptolemaic. One of these alternatives might be particularly good on planetary latitudes, another on the prediction of retrograde arcs, still another passing fair with planetary longitudes ... and so forth. The

[40] Preface to D.R.
[41] *The Copernican Revolution*, pp. 138ff.

Ptolemaic monster was hydra-headed. But, with all that, its many eyes were not very good. The predictions rarely corresponded with the observations. Indeed, centuries of use had magnified the errors, just as the small errors in a clock mechanism increase as time passes. The monster was made more unmanagable by the fact that many of the earlier observations which structured the Ptolemaic techniques, were poorly made, and inaccurately described. Errors in transcription, translation and transmission multiplied the monster's many faces. Such a beast could not be tamed, not by Ptolemy, Copernicus, Kepler, Newton or even by an Einstein. It had to be decomposed and reconstructed along more intelligible, more rational lines. The new structural anatomy which Copernicus introduces into astronomy is aptly characterized by Kuhn:

In the Copernican system it is no longer possible to shrink or expand the orbit of any planet at will, holding the others fixed. Observation for the first time can determine the order and the relative dimensions of all the planetary orbits without resort to the hypothesis of space-filling spheres.[42]

As we shall develop in a moment, the violence of the Copernican Revolution was manifested in battlefields where Copernicus himself never chose explicitly to fight. The battlefields were cosmological, philosophical, theological – and even political. But he wrote a work which was designed in technical, geometrical and mathematically-uncompromising terms. Within those terms his accomplishments are obvious enough. Leaving aside the wider conceptual commitments of Copernicus' cosmical Constitution, it is immediately obvious to any mathematically trained person, sixteenth century or twentieth century, that *De Revolutionibus* forges astronomical technique into a unified, systematic algorithm, utterly lacking in any earlier astronomical work. This was an intra-astronomical and intra-mathematical accomplishment of the first rank.[43] At the level of the scholastics' discursive disputation, Copernicus was probably not in the same league as were his logically-trained predecessors – Cusa, Oresme, Buridan, Ockham, Scotus and Grosseteste. Yet Price places Copernicus' prime contribution *there*, in philosophy and cosmology rather than in astronomy. But *De Revolutionibus*, although it may have germinated in soil furrowed by such scholastic critics, was only in a minor way part of the growth of

[42] *Op. cit.*, p. 141.
[43] Contra-Price, cf. his *Contra-Copernicus* in Clagett (1959), pp. 197ff.

that type of philosophical inquiry. As an astronomer, Copernicus wanted more accuracy in description and prediction than the *Almagest*, or any of its variants, could provide. As a mathematician and natural philosopher, Copernicus wanted more 'system' and more 'order' – more understanding and intelligibility – than the *Almagest* could provide.

But as a theologian and metaphysician? It is not clear from *De Revolutionibus* that Copernicus specifically wanted anything more than just what was necessary to fulfill his objectives as an astronomer, a mathematician, and a natural philosopher. And in this we recognize something more akin to the kind of scientist who has molded Western thought in the last two hundred years. As Kuhn says:

For Copernicus the motion of the earth was a by-product of the problem of the planets.[44]

In his theological-metaphysical-philosophical prose, Copernicus is flatly unconvincing, uninspiring and unoriginal. The First Book of *De Revolutionibus*, the introductory prose of which is the most well known, could never (by itself) have generated the Copernican Revolution – no more than did superior prose initiate a 'Cusan Revolution' an 'Oresmean Revolution', a 'Buridanean Revolution' – or an 'Aristarchean Revolution'. It was only when the First Book was read as a Summary of all the cogent mathematical argumentation to follow, that the rumblings of the Renaissance reached any real resonance. Many before Copernicus had argued That The Universe Is Spherical (I, 1), That The Earth Also Is Spherical (I, 2), How Earth, With The Water On It Forms One Sphere (I, 3), That The Motion Of The Heavenly Bodies Is Uniform, Circular, And Perpetual, Or Composed Of Circular Motions (I, 4), Whether Circular Motion Belongs To The Earth; And Concerning Its Position (I, 5), Why The Ancients Believed That The Earth Is At Rest, Like A Center, In The Middle Of The Universe (I, 7), and The Insufficiency Of These (Foregoing) Arguments, And Their Refutation (I, 8), Whether More Than One Motion Can Be Attributed To The Earth, And Of The Center Of The Universe (I, 9) – these, of course, being the chapter headings encapsulating the arguments of the First Book.

We have gone into some detail concerning the conceptual relationship of *De Revolutionibus* with later medieval renditions of the *Almagest*. We have also outlined the geometrical relationships between these great

[44] *Op. cit.*, p. 143.

works; our contention has been that (despite Price, Hall and others) the Ptolemaic technique and the Copernican construction were not geometrically equivalent. *Observational equivalence*, or operational equivalence, is the maximum that could be claimed – even granting that there is a great deal more geometrical *similarity* between the *Almagest* and *De Revolutionibus* than obtains in most cases of observational equivalence between physical theories.

Notwithstanding his insights into the structure of these two theories, Professor Derek Price is far from clear about the equivalence issue. As we saw, his argument begins by seeking to 'establish the fundamental principle of geometrical similarity which shows that angles, distances, and indeed all mathematical techniques are quite unaffected by any change from geostatic to heliostatic systems".[45] Price concludes that the two systems are 'strictly equivalent' – a conclusion which we have challenged *in extenso*. It is a complete error to suppose that Copernicus' mathematical prowess was no greater than what was required simply to take Ptolemy's geometry and 're-coordinate' it. Because of the existence of the 'formula' (Cf. p. 245 *infra*) Price thinks Copernicus' *mathematical* contributions to the history of Western thought are far less important than his cosmological ones. This is a mistaken inference based on the mistaken conclusion which Price drew from a set of sound premises. It is true, of course, that Copernicus was not the originator of any new calculational techniques – like Descartes' analytical geometry, Newton's method of fluxions, Heaviside's operational calculus, or Einstein's generalized geometries. But that would be a poor criterion for the evaluation of important mathematical contributions to the history of thought. Lagrange and Laplace were not really very original mathematicians either; but they were brilliant mathematical astronomers. The same is true of Kepler, Adams, and Leverrier. But, amongst these towering giants, Copernicus must be *primus inter pares* – a mathematical astronomer of the very first rank. Consider again his initial perplexity: his problem was not simply to save the observationally-well-founded lines of sight given epicycles, deferents, equants and resultant orbits of any sizes, velocities, ratios and shapes whatever. These were *too many* degrees of freedom, too chaotically jumbled for any real understanding of the heavens. Copernicus stripped from himself many of these freedoms –

[45] *Op. cit.*, p. 203.

just as the best geometers today will find that, for certain purposes, they must 'degeneralize' their expositions of certain traditional problems. For Copernicus the resultant planetary orbit cannot be of *any shape whatever*. Physical intelligibility requires that they be circular. The angular velocity of an epicycle center cannot be such as to generate equal angles in equal times about *any* arbitrary point whatever within the deferent. It must be equal-angles-in-equal-times about *its own* center – the center of the deferent. The variations in the apparent diameters of the celestial bodies cannot be allowed to vary through an entire counter-observational spectrum – even though 'the positions' may be saved by doing so. And, of course, after imposing upon his mathematical speculations these boundary conditions – totally absent from the medieval versions of the *Almagest* – Copernicus must accurately generate a small subset of the consequences also generable by the Ptolemaic technique – namely, the lines of sight from Earth to planet. And he *succeeds* in doing this; despite the calculational difficulties imposed and self-imposed by the boundary conditions mentioned, while at the same time welding *all* the planetary computational constructions into a single, systematic exposition – any one part of which is fundamentally determined by the other parts, with respect to their orders, magnitudes, dimensions, velocities and apparent diameters.

This is a *mathematical* achievement of the first rank. How superficial to downgrade it just because no specifically new geometrical technique emerges in the course of the argument. But this is a use of geometry within theoretical astronomy which no one in the history of the world had ever succeeded in bringing about. To describe this Price-wise as "the *Almagest* with new coordinates, plus some anti-Aristotelian cosmology" is not only conceptually inaccurate, it is historically inaccurate as well. There are other details with respect to which the thesis of Dr. Price is illuminating – especially where he indicates that, contrary to much 'received' historical opinion, the Copernican constructions were neither mathematically simpler nor observationally *more* accurate than the Ptolemaic ones. But these insights can stand well enough by themselves without the additional weight of Price's main thesis.

As a mathematically articulate, theoretical astronomer, Copernicus was to the fourteen centuries before him like what Colorado's Front Range is to the fourteen hundred miles of uninterrupted plain lying to the

East. As a discursive cosmologist, however, Copernicus' stature was much less great. These judgments are in direct contrast to those of Professor Price, who argues that the Copernican Revolution was largely initiated by the cosmological content of *De Revolutionibus*, and not by its refined mathematical approach. The latter could hardly have been the case, Price urges, because Copernicus was simply a Ptolemaic geometer 'warmed over' by a change in coordinates.

But there is yet another dimension to the appraisal of Copernicus' place in the history of astronomical thought. For, mighty mathematical astronomer that he was, he was *also* a cosmologist. Indeed, more than any of his predecessors back to the time of Aristotle, he drew together again the astronomical and the cosmological parts, the predictive and the explanatory parts, of a serious study of the heavens.

To have fused again cosmology and astronomy was a great thing; it is what makes Copernican astronomy a miracle of theoretical explanation. But *what* cosmology did Copernicus fuse with his geometrical calculations? *How* radical were his cosmological commitments, if radical at all? It is this that we must examine now.

Newton's *Principia*[46] could *not* immediately have followed *De Revolutionibus*. Newton's objectives by far outstripped the formal techniques and cosmological commitments employed by Copernicus. However, *Principia* requires for its astronomical sections little more than the groundwork Kepler laid in *De Motibus Stellae Martis*.[47]

In short, Copernicus' work can be regarded as a cosmological realignment and a brilliant geometrical tempering of machinery forged in the *Almagest*. The *Principia* is not so directly related to the *De Revolutionibus* – but it *is* to *De Motibus Stellae Martis*. The line between Ptolemy and Copernicus is geometrically and astronomically unbroken. That between Copernicus and Newton, however, is quite discontinuous, welded only by the mighty innovations of Kepler, as we shall see in detail in Book Three, Part I. The wider cultural implications of Copernicus' work need no italics. *De Revolutionibus* could never have found its place in Western thought without the philosophical preparations of e.g., Cusa and Oresme. The ultimate intellectual consequence of 1543 is nothing less than the displacement of earth, and man, from the center

[46] Sir Isaac Newton, *Philosophiae Naturalis Principia Mathematica* (London 1687).
[47] Johannes Kepler, *Astronomia Opera Omnia*, ed. Frisch (Frankfort 1858–71).

of the universe. This is a matter not to be taken lightly by Church, or State, by philosopher, or poet. As Kuhn says:

... the *De Revolutionibus* stands almost entirely within an ancient astronomical and cosmological tradition... it gave rise to a revolution that it had scarcely enunciated... In its *consequences* [it] is undoubtedly a revolutionary work.[48]

But, while profoundly exciting to the cultural historian, this comprehensive intellectual shock has not the same impact on the scholar concerned with the internal, technical developments of astronomical calculation and theorizing. As we said, the impact on technical astronomy was great – but it was not the same as its cultural effect.

In an inverted way, Andreas Osiander recognizes the distinctions between astronomy and cosmology – prediction and explanation, mathematics and physics, kinematics and dynamics – which we have traced from Aristotle, through Geminus, Thomas and the Schoolmen. In his Foreword, posthumously made to cover the text of *De Revolutionibus*, Osiander distinguished philosophical from mathematical truth, relegating Copernicus' work to the latter idiom:

... it is the astronomer's task to employ painstaking and skilled observation in collecting together the history of celestial movements, and then – because he cannot by any reasoning determine the true causes of these movements – to imagine, or construct, whatever causes or hypothese he pleases such that, by assuming these, those same movements can be calculated from the principles of geometry for the past and also for the future ... it is not necessary that these hypotheses be true ... it is enough that they provide a calculus fitting the observations ... this art is profoundly ignorant of the causes of the apparently irregular movements ... permit these new hypotheses to make a public appearance amongst older ones no more probable ...[49]

We must, with Rheticus, Kepler and Gassendi, think Osiander unfaithful to Copernicus' intentions. But one might well wish Martin Luther had seen the same distinction as clearly as Andreas:

People gave ear to an upstart astrologer who strove to show that the earth revolves, not the heavens or the firmament, the sun and the moon .... This fool wishes to reverse the entire science of astronomy.

Melanchthon concurs with Luther:

The eyes are witnesses that the heavens revolve, not in the space of twenty-four hours. But certain men, either from the love of novelty, or to make a display of ingenuity, have concluded that the earth moves, and that neither the eighth sphere nor the sun revolves....[50]

[48] *Op. cit.*, p. 134.
[49] *Loc. cit.*; Preface to the reader, my translation.
[50] Martin Luther, 'Table Talks', in A. D. White's *History of the Warfare of Science with*

Our contrast between those comprehensive philosophical explanations of the heavens, 'cosmologies', and the technical calculators which astronomers from Eudoxos to Copernicus sought in their quest for accuracy in prediction and intelligibility in explanation – this contrast must be sustained at this point.

The relative stations Copernicus and Kepler are adjudged to have in the history of astronomical thought may alter radically when viewed in turn against these distinct backgrounds. If one dwells on the philosophical-cultural implications of *De Revolutionibus Orbium Coelestium* and *De Motibus Stellae Martis*, then the orthodox account of a Copernican revolution initiated in 1543, and later reinforced by the technically industrious, but conceptually unexciting, researches of Kepler (in 1609 and 1619), seems unobjectionable. But concerning how *technical* astronomy fared as a result of these two, the appraisal cannot be reiterated unchanged.[51]

Within the history of theoretical astronomy (considered now as the evolution of finely-drawn, mathematically-articulated technical distinctions and techniques), Johannes Kepler will tower in all our later story as *the* pre-Newtonian giant. If the overstressing of Copernicus' achievements has, in most orthodox accounts, detracted from Kepler's eminence, then some compensatory understressing of the technical contents of *De Revolutionibus* may be excused here. This is just the reverse of our objectives Contra-Price – where we sought to build up the astronomical content of *De Revolutionibus* as against its cosmological content. But now our objectives are Pro-Kepler: we must therefore minimize (comparatively) Copernicus' technical contributions as these contrast with the towering astronomical achievement of Johannes Kepler.

---

*Theology in Christendom* (New York 1896), I, 126, Philip Melanchthon, *Initia Doctrinae Physicae* (Wittenberg, 1549).

[51] Similarly, to trace the intra-scientific import of special relativity requires a decision *not* to be impressed by the wider implications of this novel hypothesis. To do fundamental physics today one must study Einstein's *algebra*, not the reactions of divines, demagogues, and despondent dramatists at the turn of the century. Analogously, it is legitimate to consider Copernicus' work intra-astronomically, with respect only to the internal development of later science, and to the technical innovations of his 15th Century predecessors. This need not detract from the more publicized story about a Copernican revolution in the wider history of ideas.

We have seen that, until the XVIth century, astronomy turned on two principles: (a) the geostatic principle, and (b) the circularity principle. The grips which these principles had on astronomer's minds, however, were not equally strong.

The geostatic principle had often been challenged before Copernicus, as we have made abundantly clear. This is conceded in *De Revolutionibus* – indeed, stressed. That an idea had already figured in ancient learning minimized its cosmological shockingness for Copernicus' contemporaries. Thus Philolaus, Hiketus, Ekphantus, Aristarchus, Herakleides, Capella, Erigna, Bacon, Grosseteste, Scotus, Ockham, Albert of Saxony, Buridan, Oresme and Cusa had all seriously entertained the conception of a moving earth, whatever their ultimate attitudes towards the truth of such an hypothesis.[52]

These earlier thinkers recognized, as had Ptolemy himself, the calculational advantages of so ordering their computations. Ptolemy would have adopted outright this formally simpler and systematically more elegant device, *had the facts permitted*. Observational facts, not divine principles or fear of heresy, prevented Ptolemy from being a heliocentrist. Technical astronomers held no strong feelings against a central sun. It was just a factually false notion. To say an idea is false is not to recoil from it in horror. So far from being an inventive heretic, therefore, Copernicus was moving on familiar ground towards a more satisfactory algorithmic structure. What mattered to the astronomer was accurate and systematically-ordered descriptions and predictions. If displacing the reference system achieved this, the displacement would have been accepted in the 2nd century just as in the 16th.

Remember, not since Kalippus had astronomy been *strictly* geocentric. For Apollonios, Hipparchus, Ptolemy, and all subsequent geostatic astronomers, the earth's center was never identical with that of the universe. No astronomer ever proposed that the *Almagest* literally pictured the actual dispositions of the heavenly bodies.[53] It was a calculating device merely. Liberality could be expected with respect to its elements. It is within this liberal astronomical tradition that Copernicus' own heliocentrism should be viewed. Given the latitude to which mathe-

---

[52] Cf. Book One, Part I and Book Two, Part I.
[53] And yet, were it really geometrically *equivalent* to the Copernican constructions, the *Almagest* would have had to have been just as 'pictorial' as *De Revolutionibus*.

matically-oriented astronomers had become accustomed, Copernicus' suggestions were to be subjected to but two intra-astronomical tests: (1) accuracy in description and prediction, and (2) lucidity in intelligible systematization. A sacrifice of the geocentric principle would have seemed to theoreticians a small price to pay for these latter two commodities. And this was, indeed the outlook of Rheticus, Reinhold, Galileo, Kepler, etc.

The Circularity Principle, however, is different. In 2000 years of technical, computational astronomy it had never been questioned.[54] Copernicus did not question it; Ptolemy could not. Given the conceptual context within which ancient thought thrived, how could anyone have questioned this principle? The reasons for this complete acceptance are partly observational, partly philosophical, and strongly reinforced by other aesthetic and cultural factors.

First, the observational reasons. *The* obvious natural fact to ancient thinkers was, as we saw in Book One, the diurnal rotation of the heavens. Not only did Draco, Cepheus, and Cassiopeia spin circles around Polaris, but stars which were not circumpolar rose and set at the same place on the horizon each night. Nor did a constellation's stars vary in brightness during the course of their nocturnal flights. The conclusion was set out as our First Great Fact of the Heavens – the distances of the constellations did not vary and their paths were perfectly circular around us. Moreover, the sun's path over earth described a segment of a great circle; this was clear from the contour of the shadow traced by a gnomon before and after noon.

As early as the VIth century B.C. the earth was known to be spherical. Ships disappear hull-first and mast-last over the horizon; approaching shore their topsails appear first. Earth, being at the center of the universe, would have the same shape as the universe; so, e.g., did Aristotle argue, although this may not really be an *observational* reason in favor of circularity. The 'discoid' appearances of sun and moon were also felt to indicate the shape of celestial things.[55]

In the light of all this, one would require *special* reasons for saying that the paths of the heavenly bodies were other-than-circular. *Why*

---

[54] A possible exception to this is Cusa, Cf. J. Dreyer, *A History of Astronomy*, New York 1953, 286.
[55] Cf. Book One, Part I.

should the ancients have supposed the diurnal rotation of the heavens to be elliptical? Or oviform? Or angular? To square with what observation? There were absolutely no reasons for such suppositions then. This, conjoined with the considerations above, made the circular motions of heavenly bodies appear an almost directly observed fact.

Additional philosophical considerations, advanced notably by Aristotle [56] supported further the Circularity Principle. By distinguishing superlunary (celestial) and sublunary (terrestrial) existence, and reinforcing this with the four-element physics of Empedocles, Aristotle came to speak of the stars as perfect bodies, which moved in only a perfect way, viz. in a perfect circle.

Now what is perfect motion? As we saw, it must be motion without terminus. Because motion which begins and ends at discrete places would be incomplete. Circular motions, however, since they are eternal and perfectly continuous, lack termini. It is never motion *towards* something. Only incomplete, imperfect things move towards what they lack. Perfect, complete entities, if they move at all, cannot move towards what they lack. They move only in accordance with what is in their natures. Thus, circular motion is itself one of the essential characteristics of completely perfect, quintessential, celestial existence.

Copernicus' signal achievement was to have invented systematic astronomy. The *Almagest* and most of the *Hypotheses* outline Ptolemy's conception of his own task as the provision of computational tables, severally independent calculating devices for the prediction of future planetary perturbations. Indeed, in the Halma edition of Theon's presentation of the *Hypotheses* there is a chart setting out (under six distinct headings) otherwise unrelated diagrams for computing the planetary motions. No attempt is made by Ptolemy in the *Syntaxis* to weld into a single scheme (*à la* the Eudoxian Aristotle), these independent predicting-machines. They all have this in common: the earth is situated near the center of the deferent. But that one should superimpose all these charts, run a pin through the common point, and then scale each planetary deferent larger and smaller (to keep the epicycles from 'bumping') – this is contrary to any intention Ptolemy ever expresses. This, although the *Hypotheses* is in part an imperfect Aristotelian cosmology – crystalline spheres and all. Ptolemy might even have sup-

[56] Book One, Part I.

posed the planets to move at infinity. His problem is to forecast where, against the inverted bowl of night, some particular light will be found at future times. His problem concerns longitudes, latitudes, and angular velocities. The distances of these points of light is a problem he cannot master, not beyond crude conjectures as to the orderings of the planetary orbits viewed outward from earth. The apparent diameters of the Moon and Venus escape him. But none of this has prevented scientists, philosophers, and even historians of science, from speaking of the Ptolemaic *system*, in contrast to the Copernican system. This is a mistake, as we saw. It is engendered by confounding the Aristotelian cosmology in the *Almagest* with its geocentric astronomical calculations.

All Copernicus' planetary calculations have been seen to be *interdependent*. He cannot, e.g., compute the retrograde arc travelled by Mars, not without also making suppositions about the earth's own motion.[57] He cannot describe eclipses without entertaining some primitive form of the three-body problem. In Ptolemaic terms, however, eclipses and retrograde motion were phenomena *simpliciter*, to be explained directly as possible resultants of epicyclical combinations. In systematic astronomy retrogradations become part of the conceptual structure of the system; they are no longer a puzzling aspect of intricately variable, local planetary motions.

Another contrast stressed when discussing Ptolemaic vs. Copernican astronomy, turns on the idea of *simplicity*. It is often stated that Copernican astronomy is 'simpler' than Ptolemaic. Some even say that this is the reason for the ultimate acceptance of the former. Thus Professor Henry Margenau remarks:

A large number of unrelated epicycles was needed to explain the obervations, but otherwise the [Ptolemaic] system served well and with quantitative precision. Copernicus, by placing the sun at the center of the planetary universe, was able to reduce the number of epicycles from eighty-three to seventeen. Historical records indicate that Copernicus was unaware

[57] These are, in principle, the same suppositions one must make in defending heliocentrism at all. No one until Bessel could point to *stellar parallax*, although Bradley's 'aberration' is contingent upon the earth's motion just as is *stellar parallax*. Since Aristarchus this was the observational thorn in the heliocentric theory's flesh. But Copernicus could point to an observational simulacrum of the same phenomenon, viz., the retrograde motions of Mars and Venus. Explaining the latter parallactically was the heart of the Copernican system. So the naked-eye absence of *stellar parallax* need no longer have been deadly to the idea of a moving earth. Retrograde planetary motion, heliocentrically described, became an empirically respectable instance of exactly the same thing – parallax due to the earth's motion.

of the fundamental aspects of his so-called 'revolution', unaware perhaps of its historical importance, he rested content with having produced a *simpler* scheme for prediction. As an illustration of the principle of simplicity the heliocentric discovery has a peculiar appeal because it allows simplicity to be arithmetized; it involves a reduction in the number of epicycles from eighty-three to seventeen.[58]

Without careful qualification this can be misleading. If in any one calculation Ptolemy had to invoke 83 epicycles *all at once*, while Copernicus never required more than one-third this number, then (in the sense obvious to Margenau) Ptolemaic astronomy would be simpler than Copernican. But I know of no single planetary problem which ever required of Ptolemy more than four epicycles at one time. This, of course, results from the non-systematic, 'cellular' character of the Ptolemaic technique. Calculations within the Copernican framework always raised questions about planetary configurations *en toto*. These could be met only by considering the dynamical and kinematical elements of several planets at one time.* This is more ambitious than Ptolemy is ever required to be when he faces his isolated problems. Thus, in no ordinary sense of 'simplicity' is the Ptolemaic theory simpler than the Copernican. The latter required juggling several linked elements simultaneously. This was not simpler but much more difficult than exercises within Ptolemy's astronomy.

Analogously, anyone who argues that Einstein's theory of gravitation is simpler than Newton's must say rather more in order to explain how it is that the latter is mastered by student-physicists, while the former can be managed (with difficulty) only by accomplished experts.

In some sense, of course, Einstein's theory *is* simpler than Newton's, and there is a corresponding sense in which Copernicus' theory is simpler than Ptolemy's. But simplicity here refers to *systematic simplicity*: this is the sense in which a deductive calculus is simpler than a heap of calculational tools – the sense in which advanced algebra is simpler than bookkeeping. The number of primitive ideas in systematically-simple theories is reduced to a minimum. The axioms required to make the theoretical machinery operate are set out tersely and powerfully, so that all permissible operations within the theory can be traced rigorously back to

* *Editor's note:* The dynamical elements here are those of Copernican perfectly-circular motions, not Newton's theory of gravitation.

[58] H. Margenau, *The Nature of Physical Reality*, New York 1950, 97: actually Copernicus requires a total of 34 epicycles!

these axioms, rules, and primitive notions. In fact, the simpler $\theta$ is in this algorithmic sense, the less simple it is psychologically, and *vice versa*. This characterizes Euclid's formulation of geometry, an algorithm, but not Ptolemy's astronomy. There are in the *Almagest* no rules for determining *in advance* whether a new epicycle will be required for dealing with aberrations in lunar, solar, or planetary behavior. The strongest appeal of the Copernican formulation consisted in just this: ideally, the justification for dealing with special problems in particular ways is completely set out in the basic 'rules' of the theory. The lower-level hypotheses are never 'ad hoc', never introduced *ex post facto* just to suck up into the theory some otherwise recalcitrant datum. Copernicus, to an extent unknown by a millenium of Ptolemaic predecessors, approximated to Euclid's mighty vision. *De Revolutionibus* is not just a collection of facts and techniques. It is an organized system of these things. Solving astronomical problems requires, for Copernicus, not a random search through unrelated tables, but a regular employment of the rules defining the entire discipline. Ptolemy was a celestial contractor-engineer. He fits his tools with wit and ingenuity to each new problem as it arises. Copernicus was a celestial architect. Until he understood the structure of the entire vault he was unwilling to cement in any constituent pieces.

Hence, noting the simplicity achieved in Copernicus' formulation does not provide another reason for the acceptance of *De Revolutionibus*, another reason beyond its systematic superiority. It provides exactly the same reason.

1543 A.D. is often venerated as the birthday of the Scientific Revolution. It is really the funeral day of scholastic science. Granted, the cosmological, philosophical, and cultural reverberations initiated by the *De Revolutionibus* were felt with increasing resonance during the 300 years to follow. But, considered within technical astronomy, a different pattern can be traced.

In what does the dissatisfaction felt by Copernicus-the-astronomer consist? What in the *Almagest* draws his fire? Geocentricism *per se*? No. The formal displacement of the geocentric principle, far from being Copernicus' primary concern, was introduced only to resolve what seemed to him intolerable in orthodox astronomy, namely, the 'unphysical' triplication of centric reference-points: one center from which the planet's distances were calculated, another around which planetary velocities were computed, and still a third center (the earth) from which observations originated. This is the arrangement Copernicus finds literally monstrous: Copernicus looked backward upon inherited deficiencies. Without abandoning too much, he sought to make orthodox astronomy systematically and mechanically acceptable. He did not think he was firing the first shot of an intellectual revolution. Nor did his strictly *technical* achievements have such results.

Most of the essential elements by which we know the Copernican Revolution – easy and accurate computations of planetary position, the abolition of epicycles and eccentrics, the dissolution of spheres, the sun a star, the infinite expansion of the universe – these and many others are not to be found anywhere in Copernicus' work.[59]

Consider just how Ptolemy's astronomy offended.

Copernicus robustly attacks Ptolemy's treatment of the troublesome third lunar anomaly in *Almagest*, Book V, Ch. 5,[60] in that it introduces the movement of the moon on its epicycle as being in no way connected with the regular movement of the epicycle on its deferent. The result is that the moon is allowed to move irregularly around its epicyclic center. This Copernicus sees as a violation of the strict principle of celestial mechanics, viz., that motion in a circle (epicyclical or deferential) must always be regular with respect to the center of that circle. In *Almagest* IX, 5; XI, 9; and XII, 1, this centric triplication is generalized for all the other planets.

The appeal of the Copernican theory, therefore, rested not so much on precise determinations, but on the integrated hypothetico-deductive character of its computational machinery. An astronomer with high observational standards, but relatively undeveloped mathematical powers (e.g., Peurbach, Regiomontanus, or Brahe), might find the Copernican reformulation wholly unsatisfying. Tycho Brahe illustrates this. The intramathematical, systematic attractions of *De Revolutionibus* were lost on the noble Dane. Observational accuracy was his main criterion, with respect to which Brahe judged Copernicus to have failed.

---

[59] Kuhn, p. 134.
[60] Cf. *De Revolutionibus*, IV, 2.

# SUPPLEMENT TO SECTION ON
# COPERNICAN THEORY

Copernicus calculated with respect to the distances and periods of Mercury as follows. Consider the diagram in Figure 69a. The inside circle represents Mercury's orbit; the outer circle, the earth's. Clearly, from the earth Mercury should always be seen close to the sun 'oscillating' to either side of it. $E_1$ and $M_1$ denote the earth and Mercury when the latter is farthest to the west. The angular distance between the two bodies at this time *averages* out as about 23°.[61]

Note the triangle $SE_1M_1$. It is right-angled. Simple trigonometry gives us:

$$SM_1/SE_1 = \sin 23° \quad 0.39.$$

Consequently, Mercury's mean distance from the sun is less than that of the earth by a factor of (roughly) 2.6.

It was known that Mercury's oscillations were periodic in each 116 days. Hence in 58 days Mercury would again be detected at maximum elongation – but now to the east, rather than to the west. But then the earth, and Mercury, will occupy different positions in their respective orbits. Designate these as $E_2$ and $M_2$. The length of arc $E_1E_2$ is as follows: it is known that earth makes a solar circuit each 365.25 days. In 58 days it will cover 0.159 of its orbit – an arc of 57°. But in another 58 days Mercury will again assume maximum elongation to the west. The positions of Mercury and earth will then be designated by $M_3$ and $E_3$. Hence, in 116 days earth describes an arc $E_1E_3$, i.e., $57° + 57° = 114°$. In this time Mercury makes more than one circumsolar circuit. Hence the periods of apparent oscillation do not coincide with Mercury's orbital period. This last, however, is computable.

The above diagram shows the side $SE_1$ of triangle $SE_1M_1$ to have moved 114° to $SE_3$. Thus $SM_1$ at $SM_3$ has also moved 114°. Hence $M_1SM_3 = 114°$ so, in 116 days Mercury describes a complete circle

[61] This distance varies between 18°⁻ and 28°⁺, a fact which Copernicus accounted for by way of the epicycles.

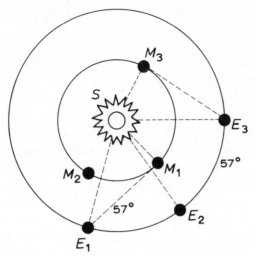

Fig. 69a.   Copernicus' computation for Mercury.

*plus* 114°, or a total arc of 474°. Mercury's orbital period, then, follows from the ratio:

$$T/116 = 360°/474°,$$
$$\therefore T = (360) \times (116)/474 \approx 88 \text{ days}.$$

Copernicus computes Venus' solar distance and its period in the same

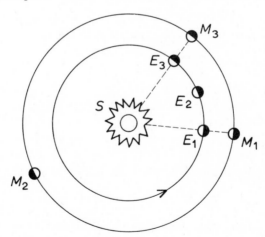

Fig. 69b.   Copernicus' computation for the superior planets.

way. Venus has an orbital period of 225 days and a mean solar distance of 0.52 that of the earth-sun distance.

Superior planets require a different treatment (see Figure 69b). Here the inside circle is the earth; the outer one is Mars'. $E_1$ and $M_1$ denote the planets when Mars and the sun are in opposition. Martian oppositions take place each 780 days. So, 390 days after opposition earth will be at $E_2$, with Mars and the sun viewed 90° from each other. In yet another 390 days, Mars will again be in opposition, with the earth at $E_3$ and Mars at $M_3$. In 780 days the earth thus sweeps out a complete circumsolar orbit, plus the arc $E_1E_3(=49°)$. The total is 769°. Meantime Mars has made one complete circuit, plus another 49°, totalling 409°. Ratios again give Copernicus the following orbital period for Mars:

$$T = (360) \times (780)/409 \approx 687 \text{ days}.$$

Copernicus uses the same method for Jupiter and Saturn, whose periods come out to 12 years and 29.5 years respectively.

What about distances? Copernicus begins calculations with the opposition point, from which he measures the angle between the sun and Mars – this is obviously 180°. This diminishes until it reaches 90° ($E'$ and $M'$, see Figure 69c). This event is observed to occur each 106 days following opposition. In that time earth has traversed an arc of

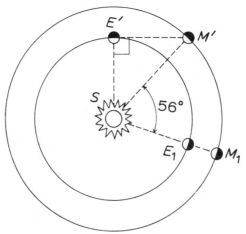

Fig. 69c.  Copernicus' computations for distances.

105°, while Mars has traversed $(360/687) \times (106) \approx 56°$. The angle $E'SM'$ is now 105° minus 56° = 49°. From $E'SM'$ we get

$$SM'/SE' = 1/\cos 49° \approx 1.5.$$

Hence Mars is 1.5 times more distant from the sun than is the earth.

In identical manner, Copernicus determines the relative distances of Jupiter and Saturn from the sun; 5 times and 9.5 times more distant, respectively.

Tycho Brahe was, as we have already seen, unflinchingly committed to the Circularity Principle. Little more needs to be said here about his unprecedented and remarkably high standards of precision in observation. Studies of his instruments, his techniques, his records and his innovations have been the staple fare for historians of astronomy for centuries. But Brahe's commitments to Circularity and to observational accuracy linked – to form one of the most interesting planetary theories of all time. Usually, Tycho's planetary theorizing is treated condescendingly by historians; he is felt to be rather a naïve geometer and an inconsistent celestial planner – as contrasted with his prowess as an observer. This will not be the position adopted here. Brahe, no less than Copernicus, must be remarked as attempting to fuse predictive astronomy and explanatory cosmology, the strict separation of which had been an identifying characteristic of celestial studies for two millenia.

In Book One, Part I, we praised Aristotle for having attempted to synthesize and systematize the most accurate and mathematically advanced astronomical techniques available in his day. Just as Einstein, with his Unified Field Theory sought to blend and unify all the unique and independent field theories of *his* day – electromagnetic, gravitational, quantum, etc. – into one algorithmically-structured system, so Aristotle (we argued) nested, dovetailed and adjusted all of the finest calculating techniques of Eudoxos and Kalippus into his Unified Cosmology. Remember, Aristotle did not sacrifice one jot of Eudoxian precision or accuracy for his systematic objectives. He had found a way of getting everything Eudoxos and Kalippus had sought, and also rendering the aggregate results 'thinkable'. A similar treatment will be tendered of Tycho here, although it is also out of step with a long tradition of standard histories of astronomy.

The utter lack of the faintest suggestion of a stellar parallax was, for

Tycho, an observational fact – a rock bottom datum not to be winked at by mathematical sophisticates. It was not to be explained away by Archimedean fantasies about the ratio of a sphere's surface to its geometrical center, the unimaginable distance of the stars, and so forth. The planets are celestial bodies; they show us what they have to show us. The 'fixed' stars are celestial bodies too; and they also show us everything they have to show us! What they show us, were we but steadfast observationalists, is that the earth does not move. They also show us that perfect circularity constitutes the shape, the structure and the principles of motion throughout the universe.

However, despite this major observational failing of the Copernican system, as well as a host of minor ones well known in the late sixteenth century, Tycho Brahe had to admit that Copernicus' manner of dealing with all the 'Sun-Dependent' features of planetary motion had a great deal theoretically to recommend it. It gave a *reason* for the fact that Mercury and Venus never achieved elongations from the sun of greater than 25° and 45° respectively: contrast the Ptolemaic and Copernican representations of this in Figure 59. This phenomenon is explained by the 'Egýptian' theory which has Mercury and Venus circling the Sun "like jewels in his crown". One can predict that Mercury will never exceed 25°, nor Venus 45°, precisely because one can explain these observed elongations *via* the hypothesis that these planets are circumsolar. The Ptolemaic account, on the other hand, cannot explain this at all: since Ptolemaic astronomers knew the observational facts as well as anyone else, they had to generate a descriptive-predictive technique which would chain these planets to the sun in all the appropriate calculations. But there was absolutely nothing in those calculations which would have been fundamentally upset had Venus at some time been observed at 50° from the sun, and Mercury at 40°. No *reasons* would have been placed in jeopardy by such a disclosure; the Ptolemaist simply would have adjusted his ratios, and perhaps added an epicycle, or two, so as adequately to describe what he had seen. The Copernican, however, simply could not have understood such an observation, just as a Newtonian could not understand levitation. He would have suspected an observer's error, or the occurrence of some unusual celestial phenomenon (like atmospheric refraction due to 'local weather') rather than calmly undertaking to redescribe the phenomenon.

There was also the fact that the centers of the Mercurial and Venusian epicycles required a period of exactly one year to move from West to East through the fixed stars back to an original point. Whereas, for the superior planets, the epicycle centers required to circle their deferents just those periods of time which we now recognize as constituting the Solar Periods of Mars, Jupiter and Saturn. Nonetheless, the period of these planets on their epicycles corresponded exactly to our Solar Year – 365.25 days.

And, of course, the stationary points and retrograde motions of all the planets were 'Sun-Dependent'. These remarkable manifestations of the Second Fact of Heavens (the Second Inequalities) occurred with the superior planets only when they were in Opposition to the Sun – and with the inferior planets only when they were in inferior conjunction with the Sun. These Second Inequalities of planetary motion, were made an architectural feature of the Copernican systematization – a fact of which Tycho Brahe was admiringly aware. But the First Inequalities of planetary motion (their obviously variable velocities during orbit) and, more importantly, the First Fact of the Heavens (its diurnal rotation) were much less satisfactorily explained by Copernicus. Indeed, as we have seen, Copernicus to some extent sacrificed Ptolemaic accuracy in dealing with the First Inequalities – and did so in order to build the Second Inequalities into the structure of his planetary theory. And, as to the observational fact of the diurnal rotation, how could something which was so obvious to the Ancients be winked at as *De Revolutionibus* does?

So, there are Brahe's Boundary conditions and his Initial conditions. Celestial motions, and the heavens itself, must be circular. And uniform! These are the boundary conditions. But there must be some directly observational reason for supposing the earth to move – and there is none, or was none very compelling in the 16th century. Stellar parallax does not exist for the naked eye. Nonetheless, the Canon of Frauenburg fired a resounding shot at the many sun-dependencies within our planetary system. These are Tycho's initial conditions. His task is to erect them all into a satisfying explanatory structure which, he felt, *De Revolutionibus* did not supply; this he sought without any sacrifice of descriptive-predictive accuracy, which again Copernicus was not wholly successful in achieving. Let Tycho speak for himself:

... according to the Copernican hypothesis, that space between Saturn and the Fixed Stars will be many times greater than the distance of the Sun from the Earth ... For otherwise the annual revolution of the Earth in the great orb, according to his speculation, will not turn out to be insensible with respect to the Eighth Sphere, as it ought ... the complete explanation ... is not given unless we are also informed within narrower limits in what part of the widest Aether, and next to which orbs of the planets, [the comet of 1577] traces its path, and by what course it accomplished this. So that this may be more correctly and intelligibly understood, I will set out my reflections ....

I considered that the old and Ptolemaic arrangement of the celestial orbs was not elegant enough, and that the assumption of so many epicycles, by which the appearances of the planets towards the Sun and the retrogradations and stations of the same, with some part of the apparent inequality, are accounted for, is superfluous; indeed, that these hypotheses sinned against the very first principles of the Art, while they allow, improperly, uniform circular motions not about [the orbit's] own centre, as it ought to be, but about another point, that is an eccentric centre which for this reason they commonly call an equant. At the same time I considered that newly introduced innovation of the great Copernicus, in these ideas resembling Aristarchus of Samos (as Archimedes shows in his Sand-Reckoner), by which he very elegantly obviates those things which occur superfluously and incongruously in the Ptolemaic system, and does not at all offend against mathematical principles. Nevertheless the body of the Earth, large, sluggish and inapt for motion is not to be disturbed by movement (especially three movements) any more than the Aethereal Lights are to be shifted, so that such ideas are opposed to physical principles and also to the authority of Holy Writ which many times confirms the stability of the Earth (as we shall discuss more fully elsewhere). Consequently I shall not speak now of the vast space between the orb of Saturn and the Eighth Sphere left utterly empty of stars by this reasoning, and of the other difficulties involved in this speculation. As (I say) I thought that both these hypotheses admitted no small absurdities, I began to ponder more deeply within myself, whether by any reasoning it was possible to discover an hypothesis, which in every respect would agree with both Mathematics and Physics, and avoid theological censure, and at the same time wholly accord with the celestial appearances. And at length almost against hope there occurred to me that arrangement of the celestial revolutions by which their order becomes most conveniently disposed, so that none of these incongruities can arise; this I will now communicate to students of celestial philosophy in a brief description.

I am of the opinion, beyond all possible doubt, that the Earth, which we inhabit, occupies the centre of the universe, according to the accepted opinions of the ancient astronomers and natural philosophers, as witnessed above by Holy Writ, and is not whirled about with an annual motion, as Copernicus wished. Yet, to speak truth, I do not agree that the centre of motion of all the orbs of the Secundum Mobile is near the Earth, as Ptolemy and the ancients believed. I judge that the celestial revolutions are so arranged that not only the lamps of the world, useful for discriminating time, but also the most remote Eighth Sphere, containing within itself all others, look to the Earth as the centre of their revolutions. I shall assert that the other circles guide the five planets about the Sun itself, as their Leader and King, and that in their courses they always observe him as the centre of their revolutions, so that the centres of the orbs which they describe around him are also revolved yearly by his motion. For I have found out that this happens not only with Venus and Mercury, on account of their small elongations from the Sun, but also with the three other superior planets. The apparent inequality of motion in these three remoter planets, including the Earth, the whole elementary world and at the same time the confines of the Moon in the vastness of their revolutions about the Sun, which the ancients accounted for by means of epicycles, and Copernicus by the annual motion of the Earth, is in this way most aptly represented through a coalescence of the centres of their spheres with the Sun in an annual revolution. For thus as suitable an opportunity is offered for the appearance of the stations and retrogradations of these planets and of their approach to and recession from the Earth and for their apparent variations in magnitude and other similar events, as either by the pretext of epicycles or by the assumption of the motion of the Earth. From all these things, when the former treatment by epicycles is understood, are deduced the lesser circuits of Venus and Mercury about the Sun itself, but not around the Earth, and the refutation of the ancient views about the disposition of epicycles above and below the Sun. Thus a manifest cause is provided why the simple motion of the Sun is necessarily involved in the motions of all five planets, in a particular and certain manner. And thus the Sun regulates the whole Harmony of the Planetary Dance in order that all the celestial appearances may subject themselves to his rule as if he were Apollo (and this was the name assigned to him by the ancients) in the midst of the Muses.

So much, indeed, for the rest of the more particular differences of the apparent inequality [of motion], which the ancients conceived to be represented by eccentrics and equants, and Copernicus by an epicycle on the circumstances of the eccentric, having the same angular velocity. These [differences] can also be represented in our hypothesis, either by a circle of a sufficient size in an eccentric orb about the Sun, or by a double circle in some concentric orb. Thus [in our system] no less than in the Copernican, all circular motions take place with respect to their own centre, since we have rejected Ptolemaic disorder. The manner of this we shall explain more particularly and fully in the work on the restoration of astronomy which (God willing) we have decided to elaborate. There we specifically discuss this hypothesis of celestial motions and shall demonstrate both that all the appearances of the planets agree perfectly among themselves and that these more correctly correspond [with our hypothesis] than with all others hitherto employed. So that this our new invention for the disposition of the celestial orbs may be better understood, we shall now exhibit its picture.[62]

Note that Tycho Brahe's system is kinematically (i.e., 'LOS') identical to Copernicus' system considered with all the original relative motions obtaining, but the earth held fast – like grasping the power take-off of a geared-down motor which revolves around the fixed take-off. Compare Professor Derek Price's earlier 'formula' for *geometrical relativity*. And compare again Figures 66, 67 and 68 – which convey in detail what the previous Figure is meant to present in bold outline.

I have, in truth, constructed a fuller explanation of the new disposition of the celestial orbs, in which are important corollaries of all the present cogitations. I shall add this near the end of the work, and there it will be shown first of all from the motions of comets, and then clearly proved, that the machine of Heaven is not a hard and impervious body stuffed full of various real spheres, as up to now has been believed by most people. It will be proved that it extends everywhere, most fluid and simple, and nowhere presents obstacles as was formerly held, the circuits of the planets being wholly free and without the labour and whirling round of any real spheres at all, being divinely governed under a given law. Whence

[62] Consult Figures 66, 67 and 68.

also it will be established that no absurdity in the arrangement of the celestial orbs follows from the fact that Mars in opposition is nearer to the Earth than the Sun itself. For in this way there is not admitted any real and incongruous penetration of the orbs (since these are really not in the sky, but are postulated solely for the sake of teaching and understanding the business) nor can the bodies of any of the planets ever run into one another nor for any reason disturb the harmony of the motions which each of them observes. So that the imaginary orbs of Mercury, Venus, and Mars are mixed with that of the Sun, and cross it, as will be more clearly and extensively declared in that place near (as I said) the colophon of the whole book, especially in our astronomical volume where we deal explicitly with these things.[63]

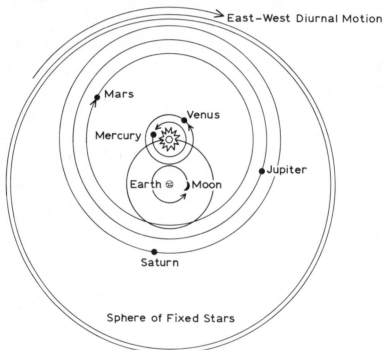

Fig. 69d.   A simplified Tychonic system.

Tycho Brahe, no less than Copernicus himself, unflinchingly held to the Principle of Perfect Circularity. In this he cannot be distinguished from a host of other post-Copernican and pre-Keplerian interim figures; Erasmus Reinhold *(Tabulae Prutelicae)*, John Field *(Ephemeris)*,

---

[63] From Ch. VIII *De Mundi Aetherie Recentioribus Phaenomenis*, in Vol. IV of Dreyer's edition, *Tychonis Brahe Dani Opera Omnia* (Copenhagen 1929) pp. 155–162. Translated by A. R. and M. B. Hall.

Robert Recorde *(Pathway to Knowledge)*, Thomas Digges *(Alae Seu Scalae Mathematicae)*, Christopher Rothmann and David Tost *(Ephemerides)*, Giovanni Benedett, Francesco Patrizio and Giordano Bruno *(De Immenso)*, and especially Michael Mästlin, the distinguished teacher of Kepler himself. All of these late sixteenth century figures were sympathetic in part, or in whole, to Copernicus' *De Revolutionibus*. And in this they, like Tycho Brahe and the young Kepler, had an almost reverent faith in the principle of Perfect Circular Celestial Motion.

Consider all of Copernicus' objections to the details of the Ptolemaic representation. And consider the natural objections to the principles of the Copernican representation, such as those set out in Brahe's passage above. Given that these objections were reasonable for the time, Tycho Brahe's systematic harmonization of the best of both worlds must be recognized for what it was – a theoretical explanation of the highest genius. What more could one ever ask from a theoretical astronomer? Brahe generated a planetary system, within which all the "sun-dependent Second Inequalities" became part of the geometrical architecture (just as in *De Revolutionibus*), but which required no fantastic inventions of the mind to account for the fact that no stellar parallax had ever been observed. The inconceivable immensity of our radial distance from the fixed stars was just such a fantastic invention. For a host of reasons, some Scriptural, some a manifestation of conservatism, and some based on his unflinching respect for the facts of observation, Tycho Brahe could not lightly let the earth move within his planetary theory. So he thus finds a way of gaining all the theoretical advantages of *De Revolutionibus* (without having to accept its observational disadvantages) and at the same time is able to utilize the observational advantages and descriptive-predictive accuracy of the *Almagest* (without having to swallow all its theoretical inelegance). The brilliant insights embodied in Brahe's achievement, its satisfactory resolution of *all* salient objections to extant astronomy, explains why the Tychonic System was almost immediately accepted, despite the complete lack of detail in its original presentation.[64]

For more than two generations after Tycho's death his system, as depicted in Figures 66, 67 and 68, counted as an acceptable alternative to

---

[64] Cf. the long quotation above.

that of Copernicus. Either was deemed suitable as a replacement for the *Almagest*. How this situation came about is in itself quite interesting. The Aristotelian distinction between superlunary and sublunary existence permeates the entire history of celestial studies, right through to the time of Copernicus. Celestial material was perfect, unchanging, incorruptible, in short – heavenly. But the brilliant and explosive supernova in Cassiopeia which appeared so dramatically in 1572 quickened Tycho's scepticism. He quickly determined that this phenomenon was superlunary It was a new star in the celestial region beyond the Moon. In this one stroke he became an opponent of orthodoxy. Tycho proved that the supernova was motionless, effecting no parallax from our point of view. It was thus superlunary. Thus the heavens changed! The spectacular comet of 1577 likewise revealed no parallax whatever. Brahe's instruments were by this time refined to a point which would invoke the respectful admiration of our Bureau of Standards even today – so his claim against the existence of parallax was given maximum weight. This being so, however, the comet must then populate the superlunary, or celestial world. What he felt to be an intra-astronomical necessity, namely, to fit the comet's actual path somewhere into the region of the planetary spheres, this is what made Brahe interrupt his long and detailed description of the comet's path to set out his general conception of the construction of the world. In the Ptolemaic technique all the space above the sun and below it is completely filled with planetary spheres – leaving thus no empty space for comets. He intuited that comets revolved above the sun, not the earth; but on either view room for their motion had to be guaranteed. If, however, all the planets (inferior and superior) revolved about the sun, and the sun with all its planetary companions then revolved about the earth, this would supply plenty of room in the region between Venus and Mars for the comets to pursue their paths.

We have noted that the great Dreyer[65] as well as Professor A. R. Hall[66] raised the spectre of equivalence between the Copernican and the Tychonic systems. Dreyer makes them 'absolutely identical' and Hall sees them as 'mathematically equivalent'. This view was objected to most strenuously. To say of two theories $\theta_1$ and $\theta_2$ that they are

---

[65] *The History of Astronomy*, p. 363.
[66] *Tycho Brahe's System of The World* (Occasional Notes of the Royal Astronomical Society, Vol. 3, 21, 1959).

*mathematically equivalent* is to say that every consequence generable from the one is also generable from the other. One cannot establish this simply by pointing out that nothing *observable* which can be inferred from the one could also be inferred from the other. For, on that criterion, the Corpuscular and Undulatory theories of light as understood in the eighteenth century – these also would have to be said to be 'mathematically equivalent'. But there are myriad geometrical consequences generable from the Tychonic system which have no counterpart whatever in the Copernican system. For one thing, Tycho's earth is fixed; Copernicus' moves. How then could two theories built on such profoundly different initial conditions be 'mathematically equivalent', or 'absolutely identical'. And consider again Tycho's orbit for Mars.[67] There is absolutely nothing in *De Revolutionibus'* theory for Mars which corresponds to such an orbit as results from the Tychonic construction. Again, the lines of sight running from Earth to Mars during a two year period (1.881 years, actually) are precisely the same in the Tychonic, the Copernican and the Ptolemaic representation, as we saw earlier. How else could they achieve *apparentias salvare?* Had any one of them failed to generate lines of sight which corresponded exactly to those actually employed in Martian observations – *that* theory would be false. And it can certainly be said that all three of these planetary theories do correspond with the facts, at least in this limited LOS-sense. All three theories, the Ptolemaic, the Copernican and the Tychonic are *observationally equivalent*. That is all – and that is all there is to it! To infer from this very well known fact to a conclusion which accords 'absolute identity' or 'mathematical equivalence' to the Copernican and Tychonic representations is logically indefensible. The situation here is precisely what was treated earlier: "the principle of geometrical relativity" will always obtain. If any one of these three representations *does* save the appearances, then each of the other two (or indeed *any* other representation which stands in a relation of geometrical relativity to the first, will also 'save the appearances'. In other words, the geometrical relativity principle, constituting as it does a sufficient condition for any theories $\theta_2$ and $\theta_3$ to 'save the appearances' if they are so related to a theory $\theta_1$ which does save the appearances – this and only this char-

[67] Compare Figure 68.

acterizes the relationships between *our* $\theta_1$, $\theta_2$ and $\theta_3$. By the nature of the constructions, one *knows* that the Tychonic and Ptolemaic constructions will be able to generate some 'appearance saving' representation *if* the Copernican constructions succeed in doing so. But this is only sufficient, it is not necessary. Some other construction, different from $\theta_1$, $\theta_2$, or $\theta_3$, might also square with the facts, despite its not standing in a relation of geometrical relativity to $\theta_1$, $\theta_2$ or $\theta_3$. And this sufficient condition is enough for us to be entitled to say *two things* of the Ptolemaic, Copernican and Tychonic constructions: we can say (1) that all three 'save the appearances', and (2) we can say of these three that the *reason* they all achieve *apparentias salvare* together is that they are intimately related *via* the Principle of Geometrical Relativity. These two things could not be said of other theories which, although they save the appearances, may not stand in this same relationship to these three theories. So Hall[68] Price,[69] Kuhn,[70] and Dreyer[71] are all correct in wishing to call attention to the interrelationship between $\theta_1$, $\theta_2$ and $\theta_3$. All three theories 'save the appearances' but also (what need not have been the case with other theory-pairs) they are related according to the Principle of Geometrical Relativity. So far, so good. But to designate this relationship as being one of 'strict equivalence', 'mathematical equivalence', 'geometrical equivalence', 'formal equivalence', or 'absolute identity' is just another illustration of where more care and attention given to strictly logical issues would help immeasurably in the attainment of clarity within history of science.

In any event, let us certainly say of Tycho Brahe that his ingenious construction has been too long underestimated. He generated the idea of a system which partook of the PGR with respect both to the Ptolemaic and Copernican constructions. Yet, he avoided what were, for the sixteenth century, the glaring factual objections to each. The result was well along the way to the achievement of the Hempelian ideal of *explanation*.

In the process, Brahe seemed to be driving the last nail into the coffin

[68] *Scientific Revolution*, Appendix, and *Occasional Notes of the Royal Astronomical Society*, Vol. 3, 1959.
[69] In Clagett, p. 203.
[70] *The Copernican Revolution*, p. 204.
[71] *History of Astronomy*, p. 363.

of late medieval Ptolemaic astronomy. Since, in his construction, the orbits of Mars and the Sun had actually to intersect in order for there to the same laws. And this, indeed, gives us the vector which drives (which is certainly the case in fact) these orbits could not be solid, crystal-line spheres. Nor was everything above the moon perfect and unchanging – as the supernova and the comet of the 1570s made clear. Indeed, at one stage, Tycho concedes tentatively that the comet might even follow an ovular rather than a perfectly circular path. The way was now clear at last for treating celestial and terrestrial phenomena as being subject to the same laws. And this, indeed, gives us the vector which drives from the great Kepler through Newton.

# BOOK THREE

## PART I

Kepler and the 'Clean' Idea

Copernicus had challenged the geocentric principle, a thing that had been done many times before him; indeed with increasing frequency during the fifteenth and sixteenth centuries. Never did Copernicus question the Circularity Principle. As we have seen, Tycho Brahe was also unflinchingly committed to that principle. Very likely, no one had ever seriously challenged the idea that all celestial objects move in perfect circles, despite a tremor to the contrary in the works of certain Oriental astronomers,[1] in some of the writings of Nicholas of Cusa, and in the astonishing concession of Tycho Brahe – wherein the paths of comets are said to be ovular. But these departures are aberrations. The young Kepler was as deeply committed to the Circularity Principle as anyone had been during the two millennia of astronomy before him. The powerful grip this principle had on astronomical thought up until the seventeenth century is illustrated by an argument in Kepler's *Mysterium Cosmographicum*:

but were hypothesis H (then being entertained by Kepler) supposed to be true, it would follow that planetary orbits would be non-circular; this is absurd, so H must be false.[2]

The property of moving in a circle was virtually a defining characteristic of anything properly known as a planet. Circularity was, indeed, as much a part of the concept *planet*, as the property of being tangible is now for us part of the concept *physical object*. Thus the suggestion of an 'intangible physical object' strikes us as conceptually untenable. Similarly, for the young Kepler (1596) the idea of a planet moving in a non-circular manner would have seemed completely absurd.

Thus it is no discredit to Copernicus that he did not question this principle. Brahe had not questioned it; Regiomontanus had not; Peurbach had not; the Merton College astronomers had not; Claudius Ptolemy had not; Hipparchus, Apollonios, Aristarchus and Euclid had not;

[1] Cf. Dreyer, *op. cit.*, Ch. XI.
[2] Kepler, *Prodromus Dissertationum Cosmographicarum Continens Mysterium Cosmographicum*, in *Johannes Kepler Gesammelte Werke*, I (Munich, 1937).

Aristotle had not; Kalippus and Eudoxos had not; Plato, and all of his predecessors, had not either. This only indicates the principle's enormous strength in the minds of astronomers. Kepler's monument, however, consists in ultimately disputing this principle, although even as late as 1600 he had no inclination to suppose anything else.

Mars had tripped up every astronomer since Eudoxos. We can easily see why. More than any other celestial light, Mars provides crucial observations for any planetary theory. The *de facto* disposition of Mars and the earth ensures that all planetary phenomena will be revealed in their purest form betwixt earth and Mars, especially just before and just after superior conjunction. Mars was a kind of 'nutcracker', either to be controlled by the 'true' theory, or otherwise to pulverize all the false ones. Only Mercury has proved to be a greater headache to students of the sky.

For Kepler, Brahe's own observations pulverize the Tychonic theory of Mars.[3] Armed with these observations of unprecedented precision, Kepler sought a better theory.

*De Motibus Stellae Martis* is a triumph of tenacity and imagination over inadequate observational and formal techniques. In the beginning, Kepler was no more disposed than had been any of his predecessors to question the circularity principle. The earlier chapters of that great work bespeak the tension between Tycho's observations and the circularity commitment. Because, submerged within Brahe's very accurate data was the solution to the "problem of empty focus" by which expression we have referred to the perennial perplexity of all pre-Keplerian astronomy, including that of Copernicus. Any attempt to reconcile data *containing* the solution (as Brahe's data did) with the circularity principle itself (which involves the negation of this solution), was bound to generate internal tensions. But the early chapters of *De Motibus Stellae Martis* do more than simply set up the problem in this way.

Even in the *Mysterium Cosmographicum* Kepler's originality and genius was apparent. He was not simply a Copernican and a Heliocentrist. He was the first Keplerian – almost as soon as he had begun to think about the planets at all. There is no need to review here his ingenious attempt to correlate the dimensions of the individual planetary orbits with the corresponding dimensions of spheres imagined to enclose nestings of

[3] Cf. Book Two, Part II.

the five regular solids; Saturn's orbit corresponded to this sphere enclosing a cube, Jupiter's to the sphere enclosing a tetrahedron, but enclosed
by Saturn's cube; Mars' orbit was a great circle enclosed by Jupiter's
tetrahedron, and also enclosing a dodecahedron; earth's orbit was enclosed by that figure, and itself enclosing an icosahedron; Venus' orbit
was a great circle of the sphere enclosed within the icosahedron, and itself
enclosing an octahedron; Mercury's orbit, finally, was the sphere just
enclosed by the octahedron.

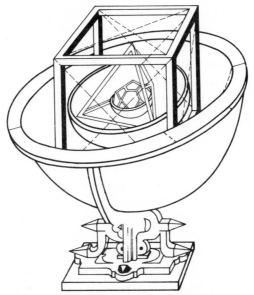

Fig. 70.   Kepler's 'Five Regular Solids' theory.

This configuration established a mathematical ratio between the
planetary orbits – of which there is no hint whatever in the work of
Copernicus, despite the fact that *De Revolutionibus* seeks to integrate
the problem of planetary distances and interrelations with the phenomenological problem involved in LOS predictions. The *Almagest* had
been concerned only with the latter. Copernicus had gone further. In his
first work, Kepler went further still. After all, Bode's Law of much later
date has no conceptual advantage over Kepler's suggestion when it
comes to determining the planetary distances.

It was in the *Mysterium Cosmographicum* also that Kepler reasoned that the sun, inasmuch as it was so near to the center of the planetary system, and so very large, must somehow be causally responsible for making the planets move as they do. Here is a conjecture of consummate importance; perhaps it is the most significant systematic hypothesis ever generated within planetary astronomy. Because, almost every astronomical study undertaken before Kepler was kinematical in conception. We have stressed the importance to Ptolemy, and to the Ptolemaists, of success in LOS predictions. Indeed, to the geocentrist *this* was the primary problem for astronomy. And it was a problem which became considerably more tractable by assuming the platform of observation to be itself fixed. The challenge then, consisted in predicting where the five 'Wanderers' were going to appear tonight, next week, next month, next year, and ten years from now. This descriptive-predictive undertaking was completely unaffected by any considerations concerned with the physical nature of these moving points of light. Even when reflected on in cosmological asides, such speculations had no connection with the predictional machinery of the *Almagest*. The celestial lights called 'Saturn', 'Jupiter', 'Mars', 'Venus', 'Mercury', the 'moon', and the 'sun' – these might have been the visible counterparts of objects as solid as diamonds on the one hand, or as diaphanous as marsh gas on the other. It did not matter. The kinematical problem remained wholly independent of what one thought about the physical constitution of these bodies. Even Rabelais' conjecture that the moon was made of green cheese would have had no bearing on the pre-Copernican problem of predicting celestial motions.

As we have seen, the Ptolemaic prediction-machinery was itself algorithmically inelegant. It was more like the tool box, or a collection of recipes, rather than a system of kinematics à la Euclid. Copernicus' triumph was to have fused Ptolemy's problem of kinematical prediction with a further hypothesis concerned with the physical geometry of the entire planetary system *de facto*. Thus he achieved two things: (1) he made predictional astronomy into a unified algorithm, and (2) he did so on the basis of heliocentrism as the unifying principle.

But this much only means that *De Revolutionibus* was a very elegant kinematical predicter. Copernicus' prowess as an astronomer was in no way matched by his intuitions as a natural philosopher. Indeed, whenever he considers how the physical geometry of the planets' motions might be connected with the nature of these bodies, his ruminations are as wooly and vague as anything to be found during the two millennia of cosmology preceding him.

It is Kepler who really begins to make astronomy into a universal physics. It is he who, in the first significant way, injects dynamics into celestial kinematics. He does not complete this undertaking – we have not yet completed it even today. It was for Newton to bring the heavens down to earth – or, to place the earth up in the highest heaven: one may choose either metaphor. It was for Newton to generate the single set of physical principles which magnificently made all of celestial kinematics, as well as all terrestrial motions, but the secondary visible effects of an understanding of the physical principles embodied within Matter, Motion, and Force. But this great unification could not even have been begun had not Johannes Kepler so clearly perceived the dynamical hiatus in all astronomical studies before his time. To encapsulate: Ptolemy's *Almagest* was a triumph in phenomenological kinematics. Copernicus' *De Revolutionibus* was a master stroke in systematic kinematics (an elegant algorithm founded deductively on geometrical axioms). Kepler's *De Motibus Stellae Martis* was the first major stroke made by celestial dynamics. His reflections initiated astrophysics, within which field Newton's *Principia* constituted our mightiest triumph.

Kepler's first physical idea, then, was monumentally anti-Aristotelian. The cause of planetary motion was not the *primum mobile*. It was the sun itself.

In his early work Kepler noted that the lines of the apsides (the line connecting perihelion and aphelion) of the Martian and the Terrestrial orbit were not parallel. Kepler felt certain that the planes of the planetary orbits must intersect somewhere, but not in the center of the ecliptic, as had been supposed by all previous astronomers, including Copernicus. He felt rather that the intersection of all these planes must reside in the physical center of the sun.[4] This assumption was clearly a physical assumption; it affected every line of Kepler's later work, and consequently all subsequent astronomy. From this single assumption Kepler determined various new methods of calculating the nodes and

---

[4] Cf. *De Motibus Stellae Martis*, Vol. 3, Chs VI and XXII.

inclinations of the orbits. These he discovered to be completely invariable.[5]

Kepler proposed that the primary problem for all astronomy should be to master the terrestrial orbit; this before ever proceeding to study of the other planets. His intuition was simply that if the earth moved, and if all observations were made from the earth, then we had to know precisely how our platform of observation was moving in order to 'reduce' all of our observations to an absolute geometry. Here, of course, was another obvious break with tradition. In geocentric theories the platform of observation was fixed in absolute space; there was no terrestrial orbit. Hence, tricky observations of complex phenomena made from a fixed platform! For axiomatic-algorithmic reasons Copernicans sought to reduce the complexities of planetary kinematics by placing several degrees of freedom within the observations themselves. Hence, tricky observations of simple phenomena made from a moving platform! Nonetheless Copernicans had given the problem of terrestrial motion only the briefest attention. Indeed, Professor Price is correct in remarking that, with respect to accuracy, the Copernican terrestrial theory offers little beyond what was available in Ptolemaic solar theory.

Kepler propounded ingenious methods for finding the earth's distance from the sun anywhere in its orbit.[6] He also introduced the principle of the bisection; assuming (a) that the terrestrial orbit is perfectly circular, and (b) that the earth's angular velocity is completely uniform, and (c) that it requires roughly 360 days to travel through those 360° – it follows that in 180 days the earth should describe a semicircle about the sun. Similarly for any larger or smaller interval of time; roughly, one day equals one degree, much as we can calculate by rule of thumb the sun's motion along the ecliptic in rough-and-ready navigational contexts today.

However, all these admirable studies forced Kepler to realize that, were the earth's orbit a perfect circle (and at this time there was no question of its being anything else), there could be no single fixed point within that orbit around which our planet could describe equal angles in equal times! This latter is a principle Kepler had already shown to be of potential explanatory value.

With this greatly improved, though still recognizably imperfect

[5] *Ibid.*, Ch. XIII.
[6] *Ibid.*, Chs XXII, XXIII.

terrestrial theory, Kepler turned his intense attention toward Mars. The immense collection of observations that had been compiled by Tycho and Longomontanus awaited him. Kepler determined Mars' distances by the same methods he had already employed in finding the earth's distances from the sun. When these were used in guiding his reasoning, however, considerable difficulties were encountered. Kepler's first Martian theory was called the 'Vicarious Theory'; it rested on one principle, fundamental to the entire astronomical tradition into which he was born: *the planet moved in a perfect circle*. This alone was proper for celestial bodies, as we have already seen *in extenso*. This alone encapsulated what Aristotle had called 'perfect motion'.[7]

So now it should be clear that in ultimately having to challenge the circularity principle Kepler was pitting his intellectual strength against one of the megaliths within the history of western scientific thought.

However, within the terms dictated by this mighty circularity assumption, the calculated distances required Mars' eccentricity to be very great indeed; so great, in fact, that the resulting equations concerning the orbit's elements were either observationally false, or mathematically inconsistent. At this stage Kepler undertook to determine the planets' orbital elements by other methods. The result was that the 'method' of equal areas in equal times – a principle which was slowly forming in Kepler's mind, and on which he was coming greatly to rely – gave errors

---

[7] Aristotle (Oxford, trans. Ross, etc.): *De Caelo*, 268b, 276b, "Circular motion is necessarily primary ... the circle is a perfect thing ... the heavens complete their circular orbit ... the heaven ... must necessarily be spherical". Also 286b, 289b. *De Generatione et Corruptione*, 337a, "Circular motion ... is the only motion which is continuous"; 338a, "It is in circular movement ... that the 'absolutely' necessary is to be found" (*Physica*, 223b, 227b, 265b, 248a, 261b, 262a). "The circle is the first, the most simple, and the most perfect figure"; [Proklus, *Commentary on Euclid's Elements* (London, T. Payne, 1788–9), definitions xv and xvi]. Cf. also Dante "Lo cerchioé perfettissima figura"; [*Convivio* (Torino, 1927), II, 13].

Cf. also Galileo, *Two Chief World Systems ...* (1632) (California, 1953), pp. 10–60, especially:

> "If such a motion [rectilinear] belonged by nature to a body, then from the beginning it would not be in its natural place; hence the ordering of the world's parts would not be a perfect one. We assume however, that the ordering of the world is perfect; consequently, it cannot by nature be intended to move in a straight line."

Cf. also Hobbes, *De Corpore Politico* (London, 1652), Part III, and M. H. Nicolson, *The Breaking of the Circle* (Evanston, 1950).

of 8′ in both excess and defect. Had the orbit really been circular this method could not possibly have given errors greater than 1′.

At this stage of his inquiry, Kepler was mildly inclined to ascribe such errors to imperfections within the method of equal areas-equal times. Very slowly, however, he came also to entertain the suspicion that perhaps his predecessors during the previous two millennia of astronomy were somewhat hasty in thinking the planetary orbits circular. Kepler's challenge to that principle seems quite natural to us after our experience with advanced conic theory, hyperbolic determinations, parabolic plots, the complexities of 'Shell Theory' within microphysics, Sommerfeld's riot of ellipses in 1916 ... etc. But this is an expression of hindsight, as mentioned in our introduction. No bolder exercise of imagination has ever been required of a natural philosopher. Indeed, Kepler, in this suspicion, dared to 'pull away the pattern'' from all the astronomical thinking there had ever been. Even the dramatic and spectacular conceptual upsets within our twentieth century required no greater break with the past. Remember, Tycho Brahe, Galileo, and many post-Keplerians never made this break at all. In Ch. XLIV of *De Motibus Stellae Martis*, Kepler begins to form three tentative arguments in support of the hypothesis that the Martian orbit might be other than circular.

(A) By supposing that Mars' orbit *was* circular, Kepler calculated the celestial longitude of the aphelion point, the eccentricity of the orbit, and the general ratio of that orbit to the terrestrial orbit. These were all totally irreconcilable with the observed elements. What is more, the result of any single combination of distances derived by these calculations was mathematically inconsistent with the results of other kinds of combinations of distances. As Kepler himself says:

Nor do the equations computed physically agree with the observed fact (the equations are represented here by the 'vicarious-hypothesis').[8]

(B) An even more significant point: a calculation of three very carefully observed distances (at 10°, 104°, and 37° of arc from aphelion). When these were plotted against a circular orbit it was revealed that the plane

---

[8] "Nec aequationes Physica computatae, observatis (quas vicaria hypothesis repraesentat) consentiebant" (p. 285).

actually 'retired within' the circular path by 350 783 and 789 parts in
100 000.

> What is to be said? ... The planet's orbit is not circular, but it recedes at either side
> slightly, and returns to the amplitude of the circle at perigee; a figure describing such a
> path is called an oval.[9]

Mars' orbit, then, may not be circular. It may consist in a curve which
coincides with the circle at the apsides, and then recedes within it; at 90°
and 270° of eccentric anomaly it will deviate most from the circle.

(C) In attempting to generate the equation for an eccentric circular
orbit by way of the method of areas (equal areas in equal times), the area
of the circle was assumed to equal the whole time of the planets' revolution. Any given sector of this circle was thus taken to be equal to the time
taken by the planet to describe that sector's particular arc. But if, in fact,
the orbit were not really a circle, errors committed through using circular sectors to measure the times taken traversing the arcs would be absolutely unavoidable. At 90° and 270°, the method of areas would give
the times as too long and the planet's motion as much too slow. But no
such errors would arise with a curve for Mars which recedes within the
circle ...

> From which it is shown what I promised to do in Chapters XX, XXIII ... that the planet's
> orbit is not a circle but has the figure of an oval.[10]

Referring to this stage in Kepler's progressing argument, Charles
Sanders Peirce says that Kepler now immediately went on to prove that
the circle was compressed. This is not true. Kepler's initial reaction to
these tentative arguments was once again to question the 'method' of
equal areas-equal times. At this stage his argument is still representable
as follows: given that the orbit *is* circular, and supposing my reasoning
to be correct, then the observations $o_1$, $o_2$, $o_3$, are directly predicted.
But $o_1$, $o_2$, and $o_3$, do not take place. *Therefore my reasoning was not
correct.* Thus, given two premises, one of which refers to the Circularity
Principle and the other to the correctness of his reasoning, Kepler is at
this stage much more ready to doubt the latter than the former.

---

[9] "Quid ergo dicendum? ... Orbitae Planetae non est circulus, sed ingrediens ad latera
utraque paulatim, iterumque ad circuli amplitudinem in perigaeo exiens. Cujusmodi
figuram itiniris ovalem appellitant" (p. 287).
[10] "Atque ex hoc quoque demonstratum, quod supra cap. xx, xxiii promisi me facturum:
Orbitam Planetae non esse circulum sed figurae ovalis" (p. 287).

After continually failing to reconcile the assumption of a circular orbit for Mars with the equations given by the methods of equal areas-equal times, he actually abandoned *the latter*! Wholly different circumstances were required in order to convince him that it was the circular orbit hypothesis which was in fact spoiling his theory. Only when the differences given to him by way of the circle were repeatedly inconsistent with those observed by Tycho, only then did Kepler begin systematically to suspect the circular orbit commitment. And even then he heads the next chapter (XLV) 'De Causis Naturalibus Hujus Deflexionis Planetae A Circulo'. However, after his inquiry under that heading he concludes:

Consider, thoughtful reader, and you will be transfixed by the force of the argument. For I could not think of any other means of making the orbit of a planet ovoid. As these things presented themselves to me in this way, the magnitude of this recession at the sides (of the circle) being securely established, as well as the agreement of the numbers, I celebrated another Martial triumph.

And even:

And we, good reader, will not indulge in this splendid triumph for more than one small day... restrained as we are by the rumors of a new rebellion, lest the fabric of our achievement perish in excessive rejoicing.[11]

Kepler now attempts to fit the non-circular curve into the framework of his other calculations; this curve coincides with the circle at the apsides. These two points are very accurately determined. But the new curve retires within the circle at 90° and 270° by 858 parts of its semi-diameter (which is supposed to contain 100 000). This is the celebrated *figuram ovalem*, whose role has been misunderstood by a great many philosophers and historians who have considered Kepler's own account.[12]

[11] "Cognita ipse lector, et vim argumenti persentices. Quia non putavi fieri ullo alio medio posse, ut Planetae orbita redderetur ovalis. Haec itaque cum ita mihi incidissent, plane securus de quantitate hujus ingressus ad latera, nimirum de consensu numerorum, jam alterum de Marte triumphum egi." And even "Ac nos, bone lector, par triumpho tam splendido dieculum unam ... indulgere, cohibitis interea novae revellionis rumoribus, ne apparatus iste nobis citra voluptatem pereat" (p. 290).

[12] In his book *The Scientific Revolution* (London, 1955) Professor A. R. Hall has clearly appreciated the significance of Kepler's first non-circular orbit hypothesis – the oviform curve. But Hall does not mark the importance of the ellipse as a mathematical tool, even in this very early phase of Kepler's work on Mars. It is an object of this part of our study to distinguish two strands in Kepler's thought: the physical hypothesis of the oviform orbit, and the mathematical hypothesis of the elliptical curve. Either hypothesis constitutes a radical break with a classical tradition. Nonetheless, even though his acceptance of non-circularity distinguishes Kepler from the centuries of Savants before him, we are nonethe-

We must now proceed carefully and in detail: these are crucial moments within the history of planetary theory.

We are obliged to recognize one important point: Kepler's first hypothesis concerning a non-circular curve for Mars was not itself an ellipse.

Whichever of these ways is used to describe the line on which the planet moves, it follows that this path, indicated by the following points, $\delta$, $\mu$, $\gamma$, $\sigma$, $\pi$, $\varrho$, $\lambda$, *is ovular, and not elliptical*; to the latter, Mechanicians wrongly give the name derived from *ovo*. The egg (ovum) can be spun on two vertices, one flatter (obtuse), one sharper (acute). Further it is bound by inclined sides. This is the figure I have created.[13]

Compare:

All of this conspires to show that the resegmentum of our eccentric circle is much larger below than above, in equal recession from the apsides. Anyone can establish this either by numerical calculation or by mechanical drawing – some eccentricity being assumed.[14]

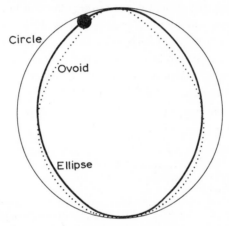

Fig. 71.   Kepler's Ovoid curve.

less obliged to distinguish within his own work two varieties of 'non-circularity'. In this very distinction it is possible to perceive another Keplerian advance almost as great as the rejection of the Circularity Principle.

[13] "Quocumque dictorum modorum delineetur linea corpus Planetae possidens, sequitur jam, viam hanc, punctis $\delta$, $\mu$, $\gamma$, $\sigma$, $\pi$, $\varrho$, $\lambda$, signatum, *vere esse ovalem, non ellipticam*, cui Mechanici nomen ab ovo ex abusu collocant. Ovum enim duobus turbinatum verticibus, altero tamen obtusiori, altero acutiori, et lateribus inclinatus cernitur. Talem figuram dico nos creasse' (p. 295, my italics).

[14] "Tot caussis concurrentibus apparet resegmentum nostri circuli eccentrici infra multo esse latius, quam supra, in aequali ab apsidibus recussu. Quod cuilibet vel numeris exploratu facile est, vel Mechanica delineatione, assumpta evidenti aliqua eccentricitate" (p. 296).

Commentators have been virtually unanimous in thinking that Kepler's first departure from the circle was immediately to an oval: that is, to a perfect ellipse.[15]

According to all these commentators Kepler's first non-circular curve was an ellipse; it happened only to be one of the wrong dimensions. In fact, that curve was not an ellipse at all. It was, rather, a *plani oviformis*.[16] There is some reason for this continual misinterpretation on the part of historians of science. Kepler was somewhat confused himself and this is partly responsible for the tradition. It must seem rather odd to any careful reader that Kepler should have jumped from the circle straight to the usually-reported ellipse without seeing at once the solution of every one of his planetary problems.

Kepler's conceptual development within *De Motibus Stellae Martis* is more intelligible when viewed in these terms. In all theories of Mars up to and including Kepler's early ones, there was but one and only one focus for the orbit. The circle, even with the sun (or the earth) placed eccentrically within it, generated a system with but one geometrical focus. Naturally, should one undertake to depart from the Principle of Perfect Circularity, the conceptual strain in doing so will be considerably less if the curve substituted is itself but a one focus curve. The ellipse is a two

[15] Cf. Small, *The Astronomical Discoveries of Johannes Kepler* (London 1804): "[Kepler] had considered the oval as a real ellipse" (p. 303). Mill: "Kepler swept all these circles away, and substituted the conception of an exact ellipse" [*A System of Logic* (8th ed.), p. 195]. Peirce: "The question is whether [the planet moves as it ought to do] ... owing to an error in the law of areas or to a compression of the orbit. He ingeniously proves that the latter is the case" (*Collected Papers*, p. 73). Wolf also obscures the important issue. He writes: "Only after trying many ovals, all larger at one end than at the other, did it occur to Kepler to try an *ellipse*, the simplest form of oval. He eventually arrived, by trial and error, at an elliptic orbit" [*A History of Science, Technology and Philosophy* (London 1952, 2nd ed.), p. 139]. For one thing, it is questionable whether an ellipse is best described as 'the simplest form of oval'. For another, as this chapter purports to make clear, to say only that the elliptical orbit was arrived at by trial and error is to obscure the important function of the ellipse even *while* the physical hypothesis of the oviform curve is dominating Kepler's thinking. Even the great Dreyer is not as lucid at this juncture as one might have wished: "For finding the areas of the oval sectors Kepler substituted for the oval an ellipse" [*History of the Planetary Systems from Thales to Kepler* (Cambridge 1906), p. 390]. But the considerations which allowed this substitution – which made it seem plausible to Kepler – are not discussed.

[16] The physical significance of the *plani oviformis* for Kepler is clear: after noting that Mars' calculated positions fall *within* the perfect circle, he straightway treats the resulting oviform as the joint-product of two physical attractive forces; that of Mars and of the sun. At this stage he never treats the ellipse in this way.

focus curve, and consequently would have constituted a considerable conceptual leap had Kepler jumped to that hypothesis 'immediately'. The oviform curve, on the other hand, *cannot* have two foci. Indeed, that is what ultimately gave Kepler all his trouble. He could work out the properties of this queer geometrical entity only by considering it as an approximation to a perfect ellipse, the properties of which had been thoroughly explored by Archimedes. Thus, only very indirectly could Kepler learn anything geometrically intelligible about this recalcitrant, one-focus figure. The idea of the oviform (one-focus) curve figured in Kepler's physical hypothesis concerning Mars' orbit. His first references to the (two-focus) ellipse figured only within the mathematical 'prop' which sustained Kepler's thinking about Mars' actual orbit. His initial observations made the physical geometry of the orbit virtually unintelligible for Kepler. But, when treated as an approximation to ("as differing but insensibly from") the geometrically-tractable Archimedean ellipse, the Martial complexities fell into an orderly array.

This move of treating observed physical phenomena as being but approximations to mathematically 'clean' conceptions developed after Kepler into what has become virtually a defining property of physical inquiry. Before his time, when the calculations seemed somewhat simpler than the facts, it was supposed that the crudity of one's observing techniques was responsible for the difference, and that *nature itself* was probably exactly as depicted by the simplest geometrical representation. After Kepler it was recognized that besides observational errors, nature itself was probably quite intricate and complicated – far in excess of what could be generated *via* the most refined mathematical techniques at any given time. Nonetheless, thinking about physical phenomena at all came to consist in 'smoothing' the experimentally-accurate descriptions to mathematically manipulable patterns. This re-orientation in methodological outlook is due primarily to Kepler's initial uses of Archimedes' ellipse in order to think about Mars' multiform motions.

Historical hindsight suggests that this might have been a fortunate accident. Suppose that the ovoid curve had been completely tractable from a mathematical point of view. Kepler might then never have gone on to introduce the ellipse into his thinking at all, not even as a geometrical prop – which was its only function at this earlier stage of *De Motibus Stellae Martis*. Kepler's own exposition indicates clearly that his having helped to introduce this 'formal crutch' made the later *physical* hypothesis of an elliptical Martian orbit (with the sun in one of the two foci) much more plausible than it ever could have been had he been dealing exclusively with one-focus curves throughout his research.[17]

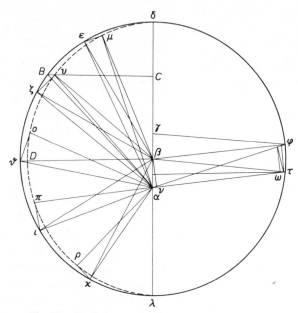

Fig. 72.    Kepler's non-circular construction (formal).

Using Archimedes' conic analyses in this way, Kepler undertook a series of fruitless attempts to find directly the quadrature of his physical Martian oviform curve. Indeed, without determining the quadrature of an orbit no equations could be forthcoming at all. He then conjectured that were the ovoid supposed to be sensibly equal to an ellipse of the same eccentricity (and described by the same greater axis), lunula cut off

[17] This diagram (*op. cit.*, p. 291) is the first non-circular curve to appear in *De Motibus Stellae Martis*, though it is hinted at on p. 276. The curve δDλ is clearly half of a perfect ellipse. But Kepler is not thinking of it here as a possible Martian orbit. The latter, as he stresses five pages later, "esse ovalem, non ellipticam" (p. 295). This is the geometrical curve to which the oviform approximates; such a device allows Kepler to enlist the help of Archimedes (p. 297). No geometer could do much with the oviform *per se*: "Ovum enim duobus turbinatum verticibus, altero tamen obturiori, altero acutiori, et lateribus inclinatus cernitur."

by it would be *but insensibly different from* that cut off by a perfect ellipse. Thus he proceeded: "If our figure were a perfect ellipse ...", and "Let us suppose then (or let it be given that) our figure is a perfect ellipse, from which it differs little. Let us see what follows therefrom."[18]

Compare: "The general geometrical properties of the perfect ellipse are manifested in the actual ovoid curve, from which it is but insensibly different, since the defects above almost exactly compensate the excesses below ..."[19]

Nonetheless, it remained impossible to find the oviform orbit's equation by the method of equal areas-equal times, which had by now regained Kepler's confidence. This, despite the exploitation of the geometrical ellipse. Kepler called on the geometers "eorumque opem imploro", a vain request when one considers that Kepler was probably the foremost geometer of his day. Still, this is not surprising. There is no *general* Cartesian expression for contours such as the *figuram ovalen* (Figure 71).

Every time this oval figure, which describes the physical path of Mars, is treated as an approximation to the pure ellipse of Archimedes, Kepler's calculations put the sun in one of the ellipses' foci. Thus "in distantia mediocri Planetae $\tau$ Sole $\alpha$" (*op. cit.*, p. 297) and "a circumferentia $\zeta$ versus Solem $\alpha$".[20] Genetically, this last remark is profoundly important.

---

[18] "Si figura nostra esset perfecta ellipsis", and "Sit autem haec figura perfecta ellipsis, parum enim differt. Videamus quid inde sequatur" (p. 297).

[19] Kepler actually puts the same point the other way round: "Concessis itaque, quae posuimus, quod planum ellipsis a plano nostri ooidis insensibiliter differat, eo quod compensatio sit inter supernos excessus ooidis supra ellipsin, et infernos defectus" (*De Motibus ...*, p. 299).

In different ways Small and Hall miss the significance of these passages. Small identifies the oval with an ellipse: "[Kepler] had considered the oval as a real ellipse" (*op. cit.*, p. 303). Hall distinguishes the oval and the ellipse so thoroughly that they seem to be unrelated in this phase of Kepler's work: "It was the accidental observation of a numerical incongruity that led [Kepler] to substitute for the oval an ellipse" (*The Scientific Revolution*, p. 125). Here, in a spectacular triumph for the history of physics, a complex, intractable phenomenon is made a subject for thought by regarding it as an approximation to a simple, easily managed mathematical entity. We must distinguish the supposed physical orbit and the mathematical tool, as Small does not. But we must not divorce them as Hall is in danger of doing.

[20] p. 299. In an obscure note Dreyer says: "The sun is not in one of the foci of this auxiliary ellipse" (*Planetary Systems*, p. 390). But he is wrong, since the quotations just given, along with the previous diagram, lack sense unless the sun is taken to reside in one of the foci of this 'auxiliary' ellipse.

Years later in Kepler's cerebrations (and one hundred pages further on in *De Motibus Stellae Martis*) Kepler comes at last to treat the ellipse, with the sun in one of the foci, as a *physical hypothesis* describing Mars' orbit. It is my conjecture that without the conceptual preparation afforded by his grappling with the oviform curve, Kepler might never have found the ellipse at all.

Figure 72 set out the first non-circular curve in *De Motibus Stellae Martis*. But the context in which it appeared was strictly calculational, and formal. The dotted line in that figure is the Archimedean ideal to which the actual physical path of Mars was supposed to approximate. Figure 73 is set in an entirely different conceptual framework.

**The Physical Elliptical Orbit for Mars**

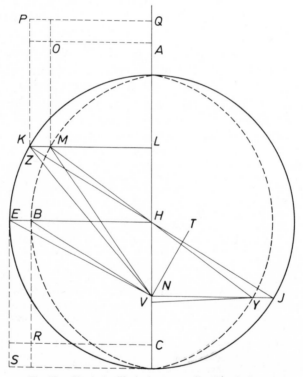

Fig. 73.   Kepler's elliptical construction (physical).

In a sense, Figure 72 is what made Figure 73 possible in Kepler's work. Midway in the march of Kepler's thoughts, the ellipse is 'there' (somewhere). But its role in the total organization of Kepler's research is now, further along, quite different. This is the geometrically simple planetary orbit which will do for Kepler's physical-observational approach to planetary theory what the heliocentric coordinates did for Copernicus' kinematical approach. Kepler's problems consisted at least in all the kinematical perplexities which had so pained Ptolemy, all these plus the systematic quest of *De Revolutionibus*; the totality of these problems had to be reconciled with the immense array of Brahe's precise observations and Kepler's own visionary intuitions of unified *physical* astronomy.

The earlier oviform theory had given the motions of Mars as being much too slow about the apsides, and as being counterfactually quick with respect to the mean distances. These were errors quite contrary to those detected in the circular orbit entertained at the beginning of *De Motibus Stellae Martis*. Kepler was uncertain whether these arose from the theory or from imperfections in the expression of the principles underlying that theory. But it appeared to him that a compromise of the circular and elliptical areas might approach the true orbit. Noting this, Kepler nonetheless tried to deduce the equations of Mars *directly* from the distances given by the oviform curve; a most loathsome undertaking, as this section of Kepler's work too readily indicates. Despite his adjustments, Mars was still described as travelling too slowly about the apsides, and much too quickly about the mean distances. Increasing the eccentricity, a move that was usually effective in many other cases, only increased the magnitude of these errors. Moreover, the oviform curve which Mars was supposed to trace had to be divided into unequal portions, because of the 'geometry' of that unprincipled figure. The curvature and length of its constituent arcs were discovered to be inversely as their distance from the sun (a prophetic correlation, as we may anticipate). However, according to the theory being generated by Kepler, only the *solar force* acts in inverse ratio of the distances; the planet's intrinsic force remains invariable.[21] This apportionment, however, is completely inconsistent with the oviform's differing curvatures and

---

[21] One can see in these reflections the influence of those Schoolmen and late medieval physicists who had so much to do with the design of the *impetus* theory.

lengths of arc.[22] This formal inconsistency led Kepler to doubt the accuracy of his measurements in the oviform orbit, construed as the joint product of the Martian and the solar force.

Kepler was presently induced to abandon any further attempts to obtain the curvature of the ovoid curve. He undertook a lengthy comparison between the distances given by the oviform and the corrected observations taken by Tycho Brahe (made at forty different points of anomaly). It soon became clear that the real orbit lay between the circle and the ovoid curve, the latter being Kepler's approximate ellipse. As he puts it himself:

> My reasoning was the same as that in Chs XLIX, L, and LVI. The circle within Ch. XLIII errs by excess. The ellipse of Chapter XLV errs by deficiency. And the excess and the deficiency are equal. There is no other middle term between a circle and an ellipse, save another ellipse. *Therefore the path of the planet is an ellipse ...* [23]

The original text has the words "ellipsis capitus XLV ..."; this suggests where and why later commentators have almost always gone wrong in their accounts of Kepler's discovery. For the curve discussed in Chapter XLV is oviform. It allows the *physical* hypothesis of the oviform orbit, which Kepler dealt with as an approximation to a formal ellipse. But now he refers to that physical hypothesis so as to indicate that the physical path of Mars *is* an ellipse, and not an intractable curve which merely approximates to an ellipse.[24] Like almost every physicist caught up in a genuine theoretical or experimental problem, Kepler may not himself have been completely clear about the strictly physical and strictly mathematical aspects of these 'frontier' hypotheses.

It thus appeared that the only orbit which could explain the data given to Kepler by Brahe itself had to be 'another' ellipse. At this critical stage

---

[22] As Galileo had argued against the impetus theorists earlier, one cannot get a variable effect from a constant cause.

[23] My italics; "Argumentatio mea talis fuit, qualis cap. xliv, 1, et lvi. Circulus cap. xliii peccat excessu, ellipsis capitus xlv peccat defectu. Et sunt excessus ille et hic defectus aequales. Inter circulum vero et ellipsin nihil mediat nisi ellipsis alia. Ergo ellipsis est Planetae iter.' p. 366. Dreyer is the only historian I know of who has seen the real meaning of this passage. He writes: "The true orbit was therefore clearly proved to be situated between the circle and the oval" (*Planetary Systems*, p. 391).

[24] Despite Kepler's misleading reference, Dreyer rightly distinguished the 'oval' and the 'auxiliary' ellipse – something which Small (cf. n. 2, p. 197) failed to do and which Hall (cf. n. 3, p. 198) completely overdoes.

in his inquiry, Kepler stumbled upon a most important correlation.[25] The 660 parts of a semi-diameter equalling 152350 were equivalent to 432 parts of a semi-diameter equalling 100000. This is nearly the number 429, which is exactly one-half of the 858 he had found to be the extreme breadth of the lunula cut off within the oval theory.[26] He focused his attention now on to the greatest optical equation of Mars (which is between 5°18′ and 5°19′). He saw that 429 was also the excess of the secant of 5°18′ above the radius 100000.[27] By way of one of the most impressive arguments in his treatise, Kepler now shows that the distances used in the circle were the *secants* of the optical equation in all points of eccentric anomaly. If, instead of these, he used the different radii to which they were the secants, the resulting calculated distances would absolutely agree with those Tycho had observed.

However, it is most unfortunate that at this point in his inquiry Kepler blundered in his calculation of the planets' positions at the determined distances – the kind of blunder every one of us makes during long manipulations with numbers. Truth seemed very reluctant to deliver herself up to Kepler, and he applied to her Virgil's allegorical couplet

Malo me Galataea petit, lasciva puella, Et fugit ad salices, et se cupit ante videri.[28]

Again he had failed. Again the data which Tycho Brahe had so laboriously compiled completely resisted the pattern Kepler proposed for the orbit of Mars. The distances his new calculations were generating, though they were observationally accurate, were formally inconsistent with the elliptical form he had ascribed to Mars' orbit. The observed distances would have required a new oviform figure again exceeding the ellipse in the first and fourth quadrants, and retiring within it in the second and third. At this point, Kepler ascribed the error to his ancillary theory of librations, wherein the planet Mars was supposed to oscillate at right angles to its orbit through its entire revolution.[29] Although this libration-theory gave very accurate distance determinations, Kepler abandoned it.

It was with little conceptual conviction that Kepler now returned to the elliptical hypothesis. This had come to seem to him to be the only

[25] Ch. LXI.
[26] Cf. *op. cit.*, p. 345.
[27] Cf. p. 346 and the accompanying diagram.
[28] Ch. LVIII, p. 364.
[29] Compare the 'librational', rectilinear trajectory resulting from uniform circular motions as set out in Figure 31 (Book One, Part II).

way of preserving his other valuable principles. He supposed that in
doing this he was appealing to something totally different from what
was involved in his theory of librations.[30] He held but the slimmest hopes
of success:

It is clear therefore that the path has cheeks; it is not an ellipse. And while an ellipse offers
justifiable equations, this cheeky figure offers unjustifiable ones.[31]

Kepler's reconsideration of the ellipse as constituting the physical
orbit for Mars, was thus something into which he retreated after finding
no other hopeful prospect of applying the principles he had by this time
so well established. The ellipse as a physical hypothesis was beginning
to beckon his attention. But now Kepler became worried nearly to
distraction through trying to *understand* why Mars should abandon (in
favor of an elliptical path) the librations – on the assumption of which
accurate calculations of distances were easily produced. He toiled on
this problem like a man possessed, as he says himself, until at last his
profound perplexities dissolved before an insight which transformed his
data, and all subsequent astronomy – and physics. In his own words:

Yet even this was not the crux of the matter. Indeed *the* great problem was this: that in considering and casting about to the very limits of my sanity, I could not discover why the
planet, to which with such great probability and agreement of observed distances the
libration LE in diameter LK could be attributed [Cf. his diagram on p. 365, *op. cit.*] why the
planet should prefer an elliptical path indicated by the equations.

Oh how ridiculous of me! As though the libration in diameter could not lead to the elliptical path. This idea carried no little persuasive force – the ellipse and the libration stand or
fall together, as will be made clear in the next Chapter. There it will be demonstrated that
there is no figure of a planet's orbit other than the perfect ellipse – the concurrence of
reasons drawn from the principles of physics, with the experiences of observations and
hypotheses being adduced in this chapter by anticipation.[32]

[30] P. 366. As Dreyer says: "This compelled Kepler to return to the ellipse, which he had
already employed as a substitute for the oval" (*Planetary Systems*, p. 392).
[31] "Patet igitur, viam buccosam esse; non igitur ellipsin. Ac cum ellipsis praebeat justas
aequationes, hanc igitur buccosam, jure ingustas praebere."
[32] "Itaque ne hic quidem valde haesi. Multo vero maximus erat scrupulus, quod pene
usque ad insaniam considerans et circumspiciens, invenire non poteram, cur Planeta, cui
tanta cum probabilitate, tanto consensu observatarum distantiarum, libratio LE [cf.
diagram on p. 365] in diametro LK tribuebatur, potius ire vellet ellipticam viam, aequati-
onibus indicibus. O me ridiculum! perinde quasi libratio in diametro, non possit esse via
ad ellipsin. Itaque non parvo mihi constitit ista notitia, juxta librationem consistere
ellipsin; ut sequenti capite patescet: ubi simul etiam demonstrabitur, nullam Planetae
relinqui figuram Orbitae, praeterquan perfecte ellipticam; conspirantibus rationibus, a
principiis Physicis, derivatus, cum experientia observationum et hypotheseos vicariae hoc
capite allegata" (*ibid.*).

Kepler's sudden insight, as contained in this passage, fused the libration theory and the elliptical theory in a manner reminiscent of what we illustrated earlier in Figures 28 and 29 (Book One, Part II). There we saw how the planetary path in question could be considered either in terms of its oscillations across the moving epicycle, or directly in terms of the resultant elliptical orbit. Kepler saw that he need no longer dwell on the rectilinear libration of Mars across an epicycle – the idea which so well approximated to the distances he had to generate. This libratory hypothesis, he saw, was geometrically equivalent to the elliptical orbit to which he had recently returned in his reflections. The Martian librations and the Martian ellipse, both of which have their independent attractions for calculation, could now be fused because they were really one and the same. The resultant coalition-conjecture had all the advantages of both.

At this stage Kepler quickly discovered his earlier arithmetic blunder. He now fully confirmed the new hypothesis.

Everything began to fall into place. Mars' orbit seemed at last to be a geometrically intelligible pattern. The elliptical areas which the planet swept over were seen to be identical. Similarly, the sums of the corresponding diametral distances were equal. The equations following from the ellipse were general expressions of the original data which Tycho had so laboriously compiled. All this made it completely clear to Kepler that Mars revolved around the sun in an ellipse, describing around that luminous body areas proportional to the planet's times.

All Kepler's labors now quickened into becoming the first *physical* planetary theory: the following three *explicantia* were the well-known result: (1) that the planetary orbits are elliptical, with the sun in their common focus (1609), (2) that the planets describe around the sun areas proportional to their times of passage (1609), and (3) that the squares of the times of the planetary revolutions are proportional to the cubes of their greater axes, or their mean distances from the sun (1619).[33] These are some of the most important principles to have been developed through the entire history of planetary theory.[34] The last mentioned, for example, carries to its consummate limit the 'systematic' idea of Copernicus; for so to have correlated the periods of the planets with

---

[33]  $T^2 \propto r^3$.

[34]  Cf. Supplementary Addendum to this Section.

their distances, Kepler was able to unite *lines-of-sight* with the physical geometry of the solar system as a whole. A record of a given planet's solar period permitted an immediate inference to its distance from the sun, and from all the other planets within our planetary system.

Kepler was soon able to generalize his Martian findings to cover all the planets. Hence, the discovery that the planetary orbit described a geometrical figure equally primitive, equally fundamental, and equally as intelligible as the circle – this constituted a finding of the greatest moment within the history of astronomy. Of course, all of Ptolemy's epicyclical and eccentric paraphernalia was immediately swept away. So was Copernicus'. The first and second inequalities of planetary motion were all now beautifully encapsulated in this Copernican-Keplerian representation: the spectacular phenomena of retrograde motions, as well as the less spectacular (but equally important) variations in the planets' apparent revolutions around the sun, these were now bracketed with Euclidean elegance in Kepler's first two Laws. Given the equal-areas in equal-times correlation, plus the conic character of Mars' path, it followed immediately that our lines of sight to that object would have to sweep to the east more rapidly at certain times of the year than at other times.

But although Kepler's position at the apex of a long history of theories of planetary motion deserves the utmost stress, his introduction of physics into astronomy – of dynamics into kinematics – is probably his most glowing achievement. His early recognition of the connection between the sun's physical properties and its dynamical effects was unprecedented. The idea of central solar forces cannot really be found in Kepler's writings – for this the world had to wait for Newton. But the idea that the sun was in some manner *causally* responsible for the motions, dispositions and appearances of the planets – this is a Keplerian innovation that cannot be doubted. To have seen that physical astronomy would require some such correlation is Kepler's triumph within planetary theory; to have failed in his conjectures concerning the nature of that physical-causal connection is relatively much less important. We cannot expect any man to be both a Copernicus and a Newton all at once. To have been a Kepler, and Kepler alone, is this man's greatest distinction.

Kepler's causal idea, at this juncture, consisted in a partial fusion, and confusion, between Gilbert's theories of magnetic effluvia and some free imagination about invisible 'spokes of force' which drove the planets forward in their paths, the hub of all those spokes residing in the center of the sun. In some difficult-to-conceive manner the 'solar spokes' were supposed to twist 'round the sun's center as invisible rays of force. Each such spoke drove a celestial object before it, thereby generating the appearances we actually encounter in the heavens. It would immediately seem to us that these spokes would have to be more than simple, smooth 'tubes' of invisible force! They would also require 'brackets' to prevent each of the planets from rocketing tangentially off into space on a straight line. Working out the details of all this was one of the triumphs of Christian Huygens, who generated a description of the Law of centrifugal force which, when artfully combined with Kepler's own third Law, paved the royal road towards Newton's monument of physical-astronomical architecture.

It is too easy simply to characterize Kepler as a 'copernican' – something which is often done in histories of astronomy. He was, of course, a heliocentrist. And his debt to Copernicus is obvious on every page. But his three great laws of planetary motion are not anticipated in any part of *De Revolutionibus*. Indeed, all of the systematic, theoretical and observational shortcomings of the Copernican conjecture are rectified in Kepler's *De Motibus Stellae Martis* and even more dramatically later in his *Epitome of Copernican Astronomy*. In Copernicus' own day there was apparently some doubt concerning whether that great man really meant for his astronomical hypothesis to be understood as affording a physical picture of the planetary system. Osiander, and many others, preferred to view the contents of *De Revolutionibus* as consisting in calculating devices merely – much in the manner of the *Almagest*. About Kepler's work, however, there can be no comparable doubt. He is clearly seeking to set out the physical geometry of our planetary system. And he wants this geometry to be such that, once perceived, it will generate in the finest detail all of the most precise and accurate observations – such as those made *in extenso* by Brahe. Moreover, the idea of physical intelligibility, and the necessity of understanding (not merely predicting) is paramount in Kepler's conjectures – all the way from his early reflections concerning the five regular solids down to his final ruminations in the *Harmonices Mundi*. The geometrical elegance, the observational accuracy, the algorithmic power and the theoretical novelty of Kepler's planetary theory was unprecedented in two millennia of heavenly studies.

And leading this remarkable synthesis was Kepler's initial rejection of the Principle of Perfect Circularity for planetary movement. The most heretical student of the heavens before his time would have been shocked at this conjecture understood as a general hypothesis; bold spirits like Brahe, Oresme, Cusa and some oriental astronomers allowed non-circular deviations in special cases. But in this man Kepler the careful student will readily perceive the *real* Revolution in the history of planetary thought. On his rejection of circularity depended every phase of Kepler's mighty astronomical synthesis. It constituted the very content of his first law. It ushered into western thought the physical thinking which turned casual star-gazing into astrophysics – more or less, depending on the seriousness of the 'star-gazing'. Even children's books on stars and planets are, these days, primers of astrophysics. And Kepler's Revolution smoothed almost every systematic rough for the ultimate physical unification brought about by Sir Isaac Newton. There have been many turning points and new departures within the history of planetary thought. None has been more dramatic, less questioned, more fraught with consequences, and less fully appreciated than the Keplerian Revolution – the main barrage within which was laid down in *De Motibus Stellae Martis*.

# SUPPLEMENTARY MATERIAL FOR BOOK THREE, PART I

*Kepler's initial determination of the terrestrial orbit*

Imagine Mars to have been observed at intervals (following opposition) equal to one, two, and then several orbital periods of that planet. It is obvious that Mars will always be at the same intersection of celestial coordinates during these observations – at the same position in its orbit. Meantime the earth will be occupying different position, since the orbital periods of earth and Mars are different. In Figure 74 $M$ = Mars and $E_0$, $E_1$, $E_2$, $E_3$, ... designates several positions of the earth.

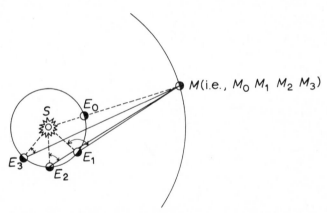

Fig. 74.   Kepler's determination of the terrestrial orbit.

Place the earth at $E_0$; this is the first observation. 687 days later (i.e., Mars' period) the earth will not have quite completed two full circuits; it will be at $E_1$. 687 days still later it will be at $E_2$ ... etc. It is quite possible thus to determine the angular differences in our successive terrestrial observations of the sun: we simply consider the apparent solar positions on the ecliptic each 687 days. In fact, the apparent 'travel' of the sun along the ecliptic in a given time will be exactly equal to the angle swept out by the line sun–earth during that same time. The ecliptic motion of the sun had long been studied before Kepler. Indeed, Kepler had before him a precise set of tables of the solar motion indicating where the sun ought to be each day. Hence angles $E_0SE_1$, $E_0SE_2$, etc., ... were very well known to Kepler. Straight observation (of $M$) then gave the angles $SE_1M$, $SE_2M$, etc. between the sun and Mars.[35] From angles $SME_1$, $SME_2$, $SME_3$, etc. within which one side $(SM)$ is constant and two angles known, Kepler then determined by trigonometry the distances $SE_1$, $SE_2$, etc. in fractions of the distance $SM$. From this one can then enter on any celestial map those points that correspond to the terrestrial positions at these different times. From which we can then draw the curve Earth traces about the sun.

Kepler initially found this terrestrial curve to be a circle within which the sun was slightly displaced from the center. Again, the ghostly 'empty focus' makes its presence (or absence) felt (see Figure 75).

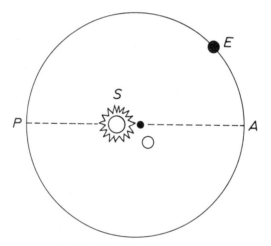

Fig. 75. The terrestrial eccentricity.

The sun was displaced approximately $\frac{1}{59}$ the circle's radius.
Kepler noticed also that the earth's orbital motion is not uniform.

---

[35] These are usually determined by measuring the positions of the sun and Mars relative to the stars, the angular distances between which are well known; Mars and the sun are rarely visible at the same time.

When closest to perihelion, $P$, it moves faster than when approaching aphelion. Kepler then drew up a table of the terrestrial motions, with the earth's positions for every day in the year set out clearly.

This much in hand, Kepler then addressed the Martian orbit directly. From all Brahe's observations, Kepler chose those which gave the planets' positions for each of its orbital periods (see Figure 76).

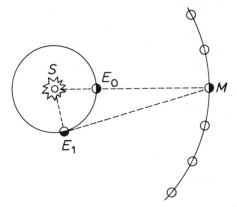

Fig. 76.   Kepler's determination of the Martian orbit.

Here $E_0$–$M$ depicts Mars at opposition. After one Martian orbital period (687 days) $E_1$–$M$ depicts the observation then made. Kepler could now determine the angle $E_1SM$ by observation. And the distance $SE_1$ came from his tables of terrestrial motion already compiled. Angle $SME_1$ was thus also determined observationally. Hence, from the triangle $SME_1$ Kepler was now enabled to find the value of $SM$ – the distance of Mars from the sun.

In precisely this manner the Martian–Solar distance at different points in the orbit became known. *De Motibus Stellae Martis* consists primarily in a sustained attempt to find the curve that will pass through all these points.

$F_1$ and $F_2$ are foci, and $O$ is the center. $AD$ is the major axis; $BE$ the minor. $AO = OD$; $BO = OE$; these are the semi-major and semi-minor axes of the ellipse. $AD$ is the line of apsides. The eccentricity of an ellipse is given by the ratio $e = OF_1/OA = OF_2/OA$ (the greater the eccentricity the more displaced are the foci from the center, and the greater the

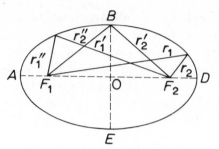

[N.B. $r_1 + r_2 = r_1' + r_2' = r_1'' + r_2'' = AD$.]

Fig. 77.   Kepler's ellipse reconsidered.

difference between the semi-major and semi-minor axes). The center's distance from the foci is as $OF_1 = OF_2 = e(OA)$. $BO$ is related to $AO$ and $e$ by:

$$BO = AO \sqrt{1 - e^2}.$$

As $BO$ increasingly differs from $AO$ the ellipse becomes more elongated. The more it thus differs from any circle. At small eccentricities the semi-major and semi-minor axes are almost identical, and the ellipse differs 'but insensibly' from a circle. Our equation indicates that if the eccentricity is 0.1 and $AO = 10$ cm (i.e., the foci are displaced from the center 1 cm) the difference between the semi-major and semi-minor is only 0.05 cm.

Kepler determined Mars' eccentricity at $\frac{1}{11}$. Its semi-major axis was 1.52 times greater than the terrestrial orbit radius. The following is such an elliptical orbit (see Figure 78).

Clearly, this figure can barely be distinguished from a circle; but the displaced focus of the sun is clearly apparent.

Mars' oscillations across the ecliptic indicated an inclination of its orbit of 2° (see Figure 79).

Near perihelion, Kepler discovered that Mars travels 37° in two months (as reckoned from the sun). Near aphelion the planet sweeps out only 25°.8 in two months. The farther a planet is from the sun, the slower it moves, so Kepler concluded. These observational data were encapsulated in the Law of Areas; the sectors swept over by Mars' radius vector are directly proportional to the times (Figure 80).

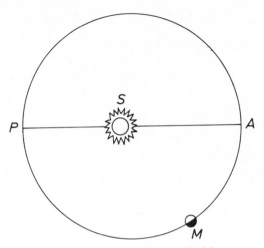

Fig. 78.   Mars' unperceivable ellipticity.

Kepler now reasoned: if the Sun must be displaced from the center of the terrestrial orbit by $1/59r$, then this should be the value of $e$. But with so small an eccentricity the semi-minor will differ from the semi-major by one part in 7000! (In the early 17th century this was observationally undetectable.) And the Second Law easily correlated with Kepler's other tables, to the effect that the earth moved faster near perihelion and slower at aphelion.

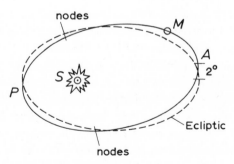

Fig. 79.   The inclination of Mars' orbital plane.

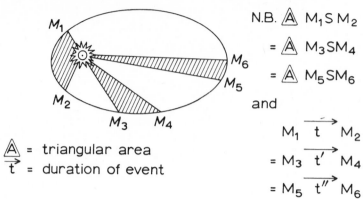

Fig. 80.   The law of Areas.

## KEPLER'S THIRD LAW

According to his data, the terrestrial-solar distance is to the Martian-solar distance as 1:1.52. The periods of revolution are as 1:1.88. Cube the first ratio and square the second: the results are 1/3.53 and 1/3.54!

That $T^2$ is proportional to $r^3$ is closely confirmed even today:

|         | Semi-major ($r$) | Period ($T$) | $r^3$ | $T^2$ |
|---------|------------------|--------------|-------|-------|
| Mercury | 0.387            | 0.241        | 0.058 | 0.058 |
| Venus   | 0.723            | 0.615        | 0.378 | 0.378 |
| Earth   | 1.000            | 1.000        | 1.000 | 1.000 |
| Mars    | 1.524            | 1.881        | 3.540 | 3.538 |
| Jupiter | 5.203            | 11.862       | 140.8 | 140.7 |
| Saturn  | 9.539            | 29.458       | 868.0 | 867.9 |

To the degree to which the last two columns are identical, to that same degree is Kepler's Third Law, $T^2 = r^3$, established.

From computations such as these, Kepler was able to compile remarkable astronomical tables which he published in 1627. These were far more comprehensive and far more accurate than anything before. In terms of his Third Law, Kepler related the distances of Jupiter's moons from the planet itself; their orbital periods were found exactly to correspond with what would have been expected in terms of $r^3 = T^2$! The terrestrial moon, however, was subject to such complex irregularities that it then proved intractable. It is incredibly difficult to cope with even now.

# SYNTHESE LIBRARY

Monographs on Epistemology, Logic, Methodology,
Philosophy of Science, Sociology of Science and of Knowledge, and on the
Mathematical Methods of Social and Behavioral Sciences

*Editors:*

DONALD DAVIDSON (Rockefeller University and Princeton University)
JAAKKO HINTIKKA (Academy of Finland and Stanford University)
GABRIËL NUCHELMANS (University of Leyden)
WESLEY C. SALMON (India University)

M. STRAUSS, *Modern Physics and Its Philosophy*. 1972, X + 297 pp.     Dfl. 80,—

SÖREN STENLUND, *Combinators, λ-terms, and Proof Theory*. 1972, 184 pp.     Dfl. 40,—

‡DONALD DAVIDSON and GILBERT HARMAN (eds.), *Semantics of Natural Language*. 1972,
X + 769 pp.     Dfl. 110,—

‡ROGER C. BUCK and ROBERT S. COHEN (eds.), *Boston Studies in the Philosophy of
Science*. Volume VIII: *PSA 1970. In Memory of Rudolf Carnap*. 1971, LXVI + 617 pp.
    Dfl. 120,—

‡STEPHAN TOULMIN and HARRY WOOLF (eds.), *Norwood Russell Hanson: What I Do Not
Believe, and Other Essays*. 1971, XII + 390 pp.     Dfl. 90,—

‡YEHOSHUA BAR-HILLEL (ed.), *Pragmatics of Natural Languages*. 1971, VII + 231 pp.
    Dfl. 50,—

‡ROBERT S. COHEN and MARX W. WARTOFSKY (eds.), *Boston Studies in the Philosophy of
Science*. Vol. VII: Milič Čapek: *Bergson and Modern Physics*. 1971, XV + 414 pp.
    Dfl. 70,—

‡CARL R. KORDIG, *The Justification of Scientific Change*. 1971, XIV + 119 pp.     Dfl. 33,—

‡JOSEPH D. SNEED, *The Logical Structure of Mathematical Physics*. 1971, XV + 311 pp.
    Dfl. 70,—

‡JEAN-LOUIS KRIVINE, *Introduction to Axiomatic Set Theory*. 1971, VI + 98 pp.     Dfl. 28,—

‡RISTO HILPINEN (ed.), *Deontic Logic: Introductory and Systematic Readings*. 1971,
VII + 182 pp.     Dfl. 45,—

‡EVERT W. BETH, *Aspects of Modern Logic*. 1970, XI + 176 pp.     Dfl. 42,—

‡PAUL WEINGARTNER and GERHARD ZECHA (eds.), *Induction, Physics, and Ethics. Proceedings and Discussions of the 1968 Salzburg Colloquium in the Philosophy of Science*.
1970, X + 382 pp.     Dfl. 65,—

‡ROLF A. EBERLE, *Nominalistic Systems*. 1970, IX + 217 pp.     Dfl. 42,—

‡JAAKKO HINTIKKA and PATRICK SUPPES, *Information and Inference*. 1970, X + 336 pp.
    Dfl. 60,—

‡KAREL LAMBERT, *Philosophical Problems in Logic. Some Recent Developments*. 1970,
VII + 176 pp.     Dfl. 38,—

‡P. V. TAVANEC (ed.), *Problems of the Logic of Scientific Knowledge*. 1969, XII + 429 pp.
    Dfl. 95,—

‡ROBERT S. COHEN and RAYMOND J. SEEGER (eds.), *Boston Studies in the Philosophy of
Science*. Volume VI: *Ernst Mach: Physicist and Philosopher*. 1970, VIII + 295 pp.
    Dfl. 38,—

‡MARSHALL SWAIN (ed.), *Induction, Acceptance, and Rational Belief.* 1970, VII + 232 pp. Dfl. 40,—

‡NICHOLAS RESCHER et al. (eds.), *Essays in Honor of Carl G. Hempel. A Tribune on the Occasion of this Sixty-Fifth Birthday.* 1969, VII + 272 pp. Dfl. 50,—

‡PATRICK SUPPES, *Studies in the Methodology and Foundations of Science. Selected Papers from 1951 to 1969.* 1969, XII + 473 pp. Dfl. 72,—

‡JAAKKO HINTIKKA, *Models for Modalities. Selected Essays.* 1969, IX + 220 pp. Dfl. 34,—

‡D. DAVIDSON and J. HINTIKKA (eds.), *Words and Objections: Essays on the Work of W. V. Quine.* 1969, VIII + 366 pp. Dfl. 48,—

‡J. W. DAVIS, D. J. HOCKNEY and W. K. WILSON (eds.), *Philosophical Logic.* 1969, VIII + 277 pp. Dfl. 45,—

‡ROBERT S. COHEN and MARX W. WARTOFSKY (eds.), *Boston Studies in the Philosophy of Science.* Volume V: *Proceedings of the Boston Colloquium for the Philosophy of Science 1966/1968.* 1969, VIII + 482 pp. Dfl. 60,—

‡ROBERT S. COHEN and MARX W. WARTOFSKY (eds.), *Boston Studies in the Philosophy of Science.* Volume IV: *proceedings of the Boston Colloquium for the Philosophy of Science 1966/1968.* 1969, VIII + 537 pp. Dfl. 72,—

‡NICHOLAS RESCHER, *Topics in Philosophical Logic.* 1968, XIV + 347 pp. Dfl. 70,—

‡GÜNTHER PATZIG, *Aristotle's Theory of the Syllogism. A Logical-Philological Study of Book A of the Prior Analytics.* 1968, XVII + 215 pp. Dfl. 48,—

‡C. D. BROAD, *Induction, Probability, and Causation. Selected Papers.* 1968, XI + 296 pp. Dfl. 54,—

‡ROBERT S. COHEN and MARX W. WARTOFSKY (eds.), *Boston Studies in the Philosophy of Science.* Volume III: *Proceedings of the Boston Colloquium for the Philosophy of Science 1964/1966.* 1967, XLIX + 489 pp. Dfl. 70,—

‡GUIDO KÜNG, *Ontology and the Logistic Analysis of Language. An Enquiry into the Contemporary Views on Universals.* 1967, XI + 210 pp. Dfl. 41,—

*EVERT W. BETH and JEAN PIAGET, *Mathematical Epistemology and Psychology.* 1966, XXII + 326 pp. Dfl. 63,—

*EVERT W. BETH, *Mathematical Thought. An Introduction to the Philosophy of Mathematics.* 1965, XII + 208 pp. Dfl. 37,—

‡PAUL LORENZEN, *Formal Logic.* 1965, VIII + 123 pp. Dfl. 26,—

‡GEORGES GURVITCH, *The Spectrum of Social Time.* 1964, XXVI + 152 pp. Dfl. 25,—

‡A. A. ZINOV'EV, *Philosophical Problems of Many-Valued Logic.* 1963, XIV + 155 pp. Dfl. 32,—

‡MARX W. WARTOFSKY (ed.), *Boston Studies in the Philosophy of Science.* Volume I: *Proceedings of the Boston Colloquium for the Philosophy of Science, 1961–1962.* 1963, VIII + 212 pp. Dfl. 26,50

‡B. H. KAZEMIER and D. VUYSJE (eds.), *Logic and Language. Studies dedicated to Professor Rudolf Carnap on the Occasion of his Seventieth Birthday.* 1962, VI + 256 pp. Dfl. 35,—

*EVERT W. BETH, *Formal Methods. An Introduction to Symbolic Logic and to the Study of Effective Operations in Arithmetic and Logic.* 1962, XIV + 170 pp. Dfl. 35,—

*HANS FREUDENTHAL (ed.), *The Concept and the Role of the Model in Mathematics and Natural and Social Sciences. Proceedings of a Colloquium held at Utrecht, The Netherlands, January 1960.* 1961, VI + 194 pp. Dfl. 34,—

‡P. L. R. GUIRAUD, *Problèmes et méthodes de la statistique linguistique.* 1960, VI + 146 pp. Dfl. 28,—

*J. M. BOCHEŃSKI, *A Precis of Mathematical Logic.* 1959, X + 100 pp. Dfl. 23,—

# SYNTHESE HISTORICAL LIBRARY

Text and Studies
in the History of Logic and Philosophy

*Editors:*

N. KRETZMANN (Cornell University)
G. NUCHELMANS (University of Leyden)
L. M. DE RIJK (University of Leyden)

‡LEWIS WHITE BECK (ed.), *Proceedings of the Third International Kant Congress, held at the University of Rochester, March 30–April 4, 1970.* 1972, XI + 718 pp.　　Dfl. 160,—
‡KARL WOLF and PAUL WEINGARTNER (eds.), *Ernst Mally: Logische Schriften.* 1971, X + 340 pp.　　Dfl. 80,—
‡LEROY E. LOEMKER (eds.), *Gottfried Wilhelm Leibnitz: Philosophical Papers and Letters.* A Selection Translated and Edited, with an Introduction. 1969, XII + 736 pp.
　　Dfl. 125,—
‡M. T. BEONIO-BROCCHIERI FUMAGALLI, *The Logic of Abelard.* Translated from the Italian. 1069, IX + 101 pp.　　Dfl. 27,—

Sole Distributors in the U.S.A. and Canada:
*GORDON & BREACH, INC., 440 Park Avenue South, New York, N.Y. 10016
‡HUMANITIES PRESS, INC., 303 Park Avenue South, New York, N.Y. 10010